WORKSHEETS
FOR CLASSROOM OR LAB PRACTICE

CHRISTINE VERITY

MATHEMATICS IN ACTION: ALGEBRAIC, GRAPHICAL, AND TRIGONOMETRIC PROBLEM SOLVING
FOURTH EDITION

The Consortium for Foundation Mathematics

Boston Columbus Indianapolis New York San Francisco Upper Saddle River
Amsterdam Cape Town Dubai London Madrid Milan Munich Paris Montreal Toronto
Delhi Mexico City Sao Paulo Sydney Hong Kong Seoul Singapore Taipei Tokyo

Reproduced by Pearson from electronic files supplied by the author.

Publishing as Pearson, 75 Arlington Street, Boston, MA 02116.

ISBN-13: 978-0-321-73835-6
ISBN-10: 0-321-73835-7

1 2 3 4 5 6 OPM 15 14 13 12 11

www.pearsonhighered.com

Table of Contents

Name: Date:
Instructor: Section:

Chapter 1 FUNCTION SENSE

Activity 1.1

Learning Objectives
1. Identify input and output in situations involving two variable quantities.
2. Identify a functional relationship between two variables.
3. Identify the independent and dependent variables.
4. Use a table to numerically represent a functional relationship between two variables.
5. Write a function using function notation.

Key Terms

Use the vocabulary terms listed below to complete each statement in Exercises 1–5.

variable **function** **independent** **dependent** **ordered pair**

1. A(n) ____________________ of numbers consists of two numbers written in the form (input, output).

2. If the relationship between two variables is a function, the output variable is called the ____________________ variable.

3. A(n) ____________________, usually represented by a letter, is a quality that may change in value from one particular instance to another.

4. If the relationship between two variables is a function, the input variable is called the ____________________ variable.

5. A(n) ____________________ is a correspondence between an input variable and an output variable that assigns a single output value to each input value.

Practice Exercises

For #6-9, use the function $y = h(x)$.

6. Determine the input. **7.** Determine the output.

6. ____________________

7. ____________________

8. Determine the function name. **9.** Write in words the equation as you would say it.

8. ____________________

9. ____________________

____

For #10-13, use the function $g(7) = 6.931$.

10. Determine the input.

11. Determine the output.

10. ______________

11. ______________

12. Determine the function name.

13. Write in words the equation as you would say it.

12. ______________

13. ______________

____

For #14-17, use the function $762 = f(t)$.

14. Determine the input.

15. Determine the output.

14. ______________

15. ______________

16. Determine the function name.

17. Write in words the equation as you would say it.

16. ______________

17. ______________

____

For #18-21, use the function $\text{salary} = s(\text{hours})$.

18. Determine the input.

19. Determine the output.

18. ______________

19. ______________

20. Determine the function name.

21. Write in words the equation as you would say it.

20. ______________

21. ______________

____

Name: Date:
Instructor: Section:

For #22-23, the input of a function C is price. The output is commission.

22. Write the function.

23. Write $C(6000) = 20$ as an ordered pair.

22. ______________

23. ______________

For #24-25, use the pairs (3, 6), (–4, 11), (16, 0), and (4, 8).

24. Do the ordered pairs represent a function?

25. Explain your answer to Exercise #24.

24. ______________

25. ______________

____

For #26-27, use the pairs (2, 7), (3, 8), (2, 9), and (4, 3).

26. Do the ordered pairs represent a function?

27. Explain your answer to Exercise #26.

26. ______________

27. ______________

____

28. For (9, 8), (9, 6), and (9, 11), explain why these pairs do not represent a function.

28. ______________

____

Concept Connections

29. Suppose that an input is the number of hours you worked at a job. Give an example of an output for this function.

____

____

30. You and three friends work at a job and each of you receive a different hourly wage, based on experience. Does this scenario, where the input is the number of hours worked, and the output is the wages received, describe a function? Why or why not?

____

____

Name: Date:
Instructor: Section:

Chapter 1 FUNCTION SENSE

Activity 1.2

Learning Objectives

1. Determine the equation (symbolic representation) that defines a function.
2. Determine the domain and range of a function.
3. Identify the independent and the dependent variables of a function.

Key Terms

Use the vocabulary terms listed below to complete each statement in Exercises 1–2.

dependent **independent**

1. ____________ variable is another name for the output variable of a function.

2. ____________ variable is another name for the input variable of a function.

Practice Exercises

For #3-5, use the function $f(x) = 5x - 6$.

3. Determine $f(3)$.

4. Determine $f(-2.7)$.

3. ____________

4. ____________

5. Determine $f(c)$.

5. ____________

For #6-8, use the function $g(y) = -8y^2 + 6.2y + 13$.

6. Determine $g(4)$.

7. Determine $g(-5.1)$.

6. ____________

7. ____________

8. Determine $g(b)$.

8. ____________

For #9-11, use the function $h(x) = 11$.

9. Determine $h(6)$.

10. Determine $h(-14.7)$.

9. ______________

10. ______________

11. Determine $h(d)$.

11. ______________

For #12-14, use the function $p(x) = \frac{5}{x}$.

12. Determine $p(-2)$.

13. Determine $p(0.5)$.

12. ______________

13. ______________

14. Determine $p(a)$.

14. ______________

For #15-17, use the function $r(x) = 4 - 2.3x$.

15. Determine $r(-7)$.

16. Determine $r(8.4)$.

15. ______________

16. ______________

17. Determine $r(c)$.

17. ______________

For #18-27, use the following scenario.

Your job requires you to attend meetings at other campus locations which are within 50 miles. You are reimbursed at the rate of $0.51 per mile for this travel.

18. Write a verbal statement that describes how the amount of reimbursement is determined.

19. Identify the input variable of the function from Exercise #18.

18. ______________

____

19. ______________

Name:
Instructor:

Date:
Section:

20. Identify the output variable of the function from Exercise #18.

21. Write the verbal statement from Exercise #18, using function notation for the input variable. Let m represent the input variable. Let R represent the function and $R(m)$ the output variable.

20. ________________

21. ________________

22. From Exercise #21, identify the dependent variable.

23. Use the equation from Exercise #21 to determine the reimbursement for travel of 74 miles.

22. ________________

23. ________________

24. Determine the domain of the function from Exercise #21.

25. Determine the range of the function from Exercise #21.

24. ________________

25. ________________

26. Determine the practical domain of the function from Exercise #21.

27. Determine the practical range of the function from Exercise #21.

26. ________________

27. ________________

28. For the function $\{(-3, 6), (9, 0), (7, 4), (4, 17)\}$, determine the domain and range.

28. ________________

____

Concept Connections

29. Explain the difference between the domain and the practical domain of a function.

____

____

30. What are real numbers?

____

____

Name: Date:
Instructor: Section:

Chapter 1 FUNCTION SENSE

Activity 1.3

Learning Objectives

1. Represent a function verbally, symbolically, numerically, and graphically.
2. Distinguish between a discrete function and a continuous function.
3. Graph a function using technology.

Key Terms

Use the vocabulary terms listed below to complete each statement in Exercises 1–3.

continuous **graphically** **discrete**

1. Functions are ____________________ if they are defined only at isolated input values and do not make sense or are not defined for input values between those values.

2. Functions are ____________________ if they are defined for all input values, and if there are no gaps between any consecutive input values.

3. When a function is defined ____________________, the input variable will be represented on the horizontal axis and the output on the vertical axis.

Practice Exercises

For #4-7, using the standard window of a graphing calculator, sketch a graph of each quadratic function.

4. $y = 0.5x^2$

5. $y = -0.5x^2$

6. $y = 0.5x^2 + 2$

7. $y = 0.5x^2 - 2$

For #8-14, use the following scenario.

The week before final exams, the test center at a community college administered make-up tests to students as follows:

Day	1	2	3	4	5
Number of tests	44	61	59	82	98

8. Plot each ordered pair as a point on an appropriately scaled and labeled set of coordinate axes.

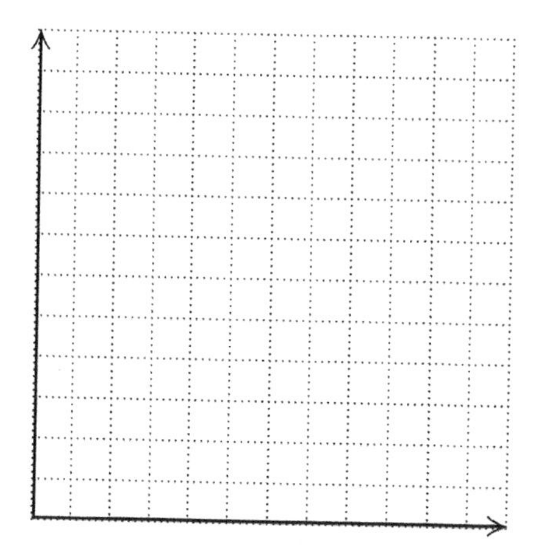

9. Determine the practical domain of the function.

10. Determine the practical range of the function.

9. ______________

10. ______________

11. Is this function discrete?

12. Explain your answer to Exercise #12.

11. ______________

12. ______________

____

13. Can this function be defined symbolically?

14. Explain your answer to Exercise #14.

13. ______________

14. ______________

____

Name: Date:
Instructor: Section:

For #15-20, use the following scenario.
At an amusement park there is a 25% employee discount for food.

15. Give a statement definition of the function.

16. Give a symbolic definition of the discount function.

15. ______________

____

16. ______________

____

17. Give a numerical definition.

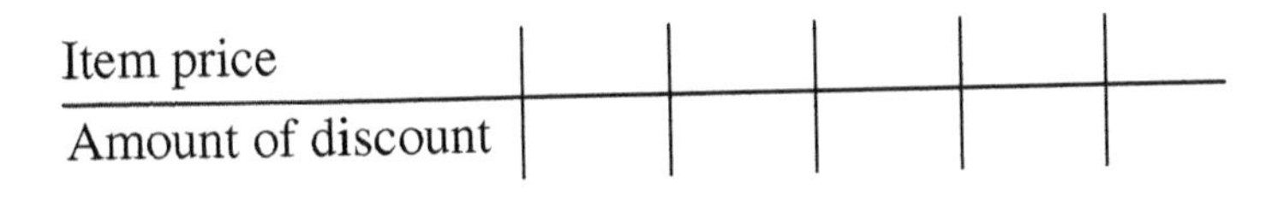

Item price					
Amount of discount					

18. Give a graphical definition.

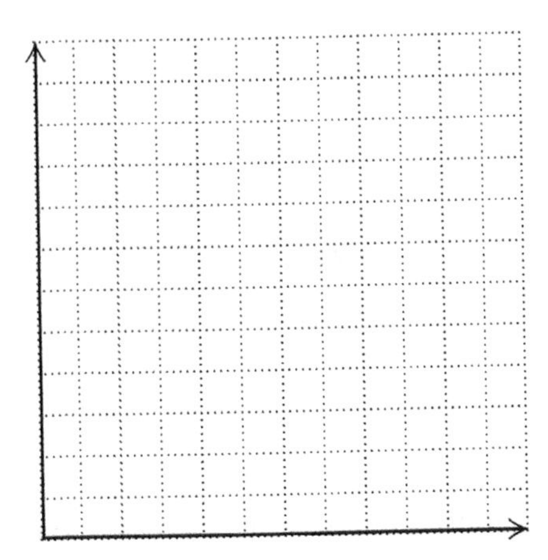

19. Does the graph of the function consist of the five points from Exercise #18?

20. Explain your answer to Exercise #19.

19. ______________

20. ______________

____

For #21-24, use the quadratic function $y = 0.0005x^2$.

21. Using the standard window of your graphing calculator to sketch a graph of the function.

22. Use the table feature of your graphing calculator to complete the following table.

x	-3	-2	-1	0	1	2	3
y							

23. Describe how you would use the results in Exercise #22 to help select an appropriate viewing window.

____

____

24. Sketch a graph of the function with the new viewing window.

For #25-28, use the quadratic function $y = 1000x^2$.

25. Using the standard window of your graphing calculator to sketch a graph of the function.

26. Use the table feature of your graphing calculator to complete the following table.

x	-3	-2	-1	0	1	2	3
y							

27. Describe how you would use the results in Exercise #26 to help select an appropriate viewing window.

____

____

28. Sketch a graph of the function with the new viewing window.

Name: Date:
Instructor: Section:

Concept Connections

29. What is the difference between a discrete and continuous function?

____

____

30. In what four ways can a function be represented?

____

____

Name:
Instructor:

Date:
Section:

Chapter 1 FUNCTION SENSE

Activity 1.4

Learning Objectives
1. Use a function as a mathematical model.
2. Determine when a function is increasing, decreasing, or constant.
3. Use the vertical line test to determine if a graph represents a function.

Key Terms
Use the vocabulary terms listed below to complete each statement in Exercises 1–4.

constant **decreasing** **increasing** **mathematical model**

1. A function is ________________ if its graph goes up to the right.

2. A(n) ________________ is an equation or a graph that fits or approximates the actual data.

3. A function is ________________ if its graph is horizontal.

4. A function is ________________ if its graph goes down to the right.

Practice Exercises
For #5-13, use the following scenario.
The value of a lake front vacation home appreciates over time. You purchase a small home for $85,000 and the value increases by $1250 per year.

5. State a question you might want to answer in this situation.

6. What two variables are involved in this problem?

7. Which variable can best be designated as the dependent variable?

8. Which variable can best be designated as the independent variable?

5. ________________

____

6. ________________

____

7. ________________

____

8. ________________

____

9. Complete the following table:

Independent Variable	1	2	3	4
Dependent Variable				

10. State in words the relationship between the independent and dependent variables.

11. Use appropriate letters to represent the variables involved.

10. ________________

____

11. ________________

____

12. Translate the written statement from Exercise #11 as an equation.

13. If you plan to keep the home for 8 years, determine the value of the home at the end of this period.

12. ________________

13. ________________

For 14-16, use the following function $f(x) = 4x - 6$.

14. Use your graphing calculator to graph the function. Make a sketch below.

15. Determine if the function is increasing or decreasing or constant.

16. Explain your answer to Exercise #15.

15. ________________

16. ________________

____

Name: Date:
Instructor: Section:

For 17-19, use the following function $g(x) = -3$.

17. Use your graphing calculator to graph the function. Make a sketch below.

18. Determine if the function is increasing or decreasing or constant.

19. Explain your answer to Exercise #18.

18. ______________

19. ______________

____

For 20-21, use the following function $f(x) = 4 - 3x$.

20. Use your graphing calculator to graph the function. Make a sketch below.

21. Determine if the function is increasing or decreasing or constant.

21. ______________

For #22–28, use the vertical line test to determine whether each graph represents a function.

22.

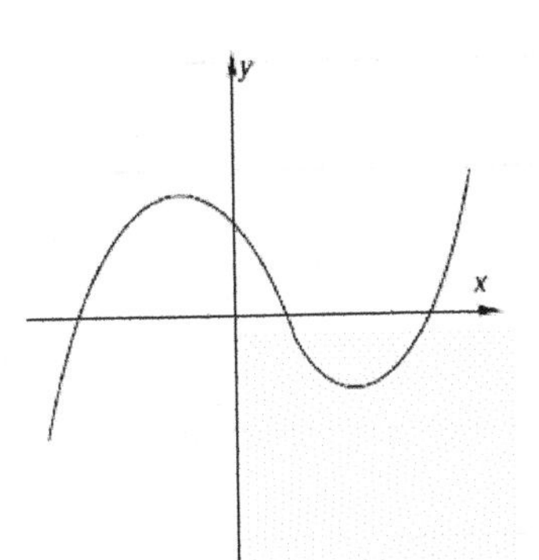

23.

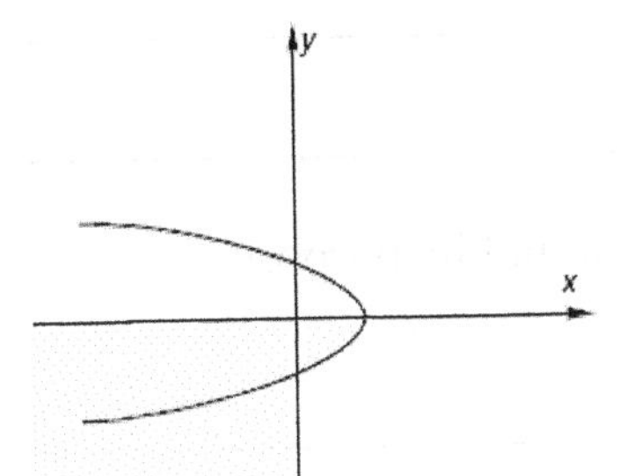

22. ______________

23. ______________

24.

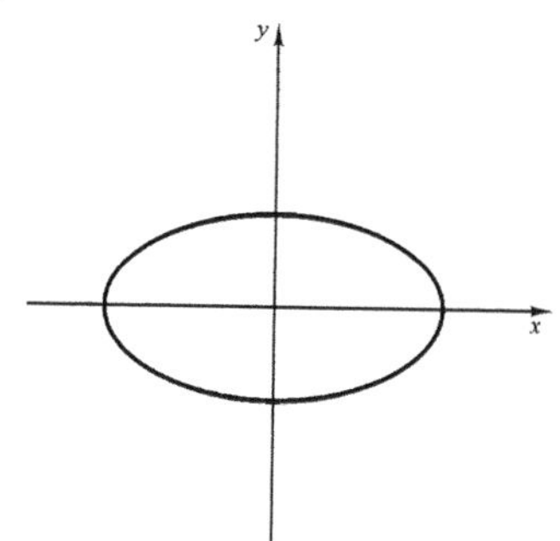

25.

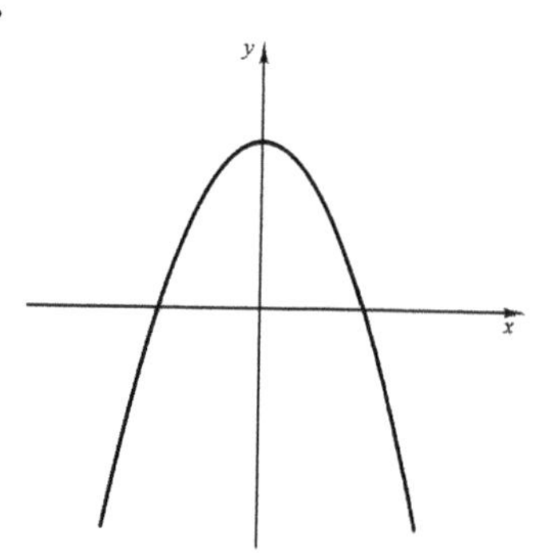

26.

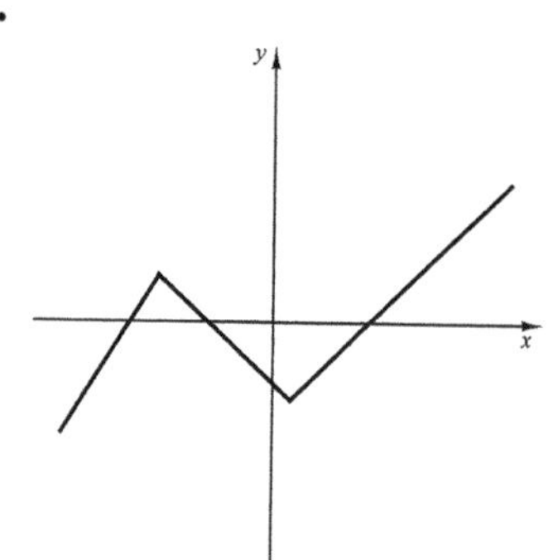

27.

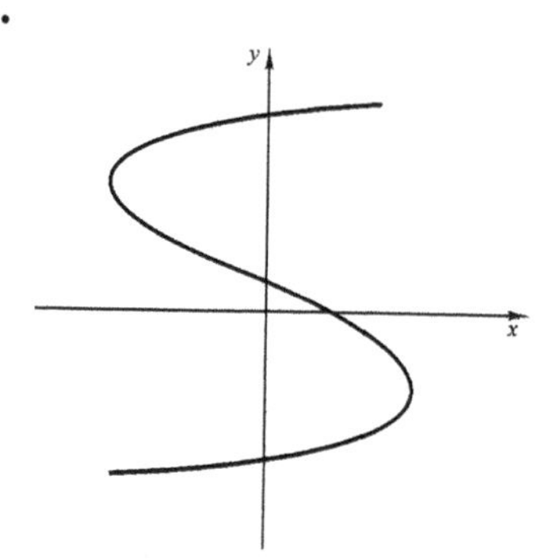

28.

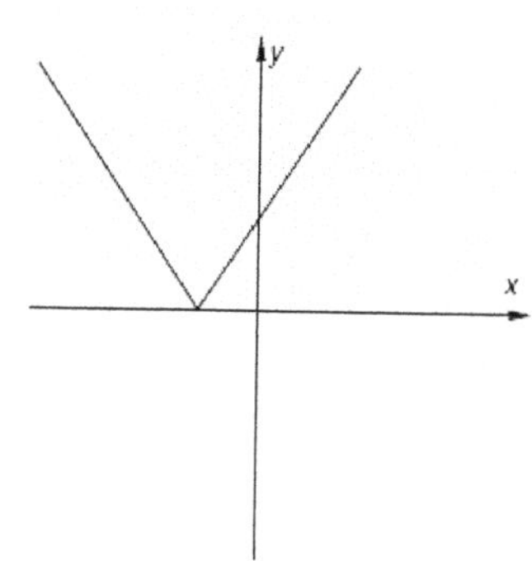

24. ______________

25. ______________

26. ______________

27. ______________

28. ______________

Concept Connections

29. What is the purpose of a mathematical model?

____

____

30. Explain the vertical line test and its purpose.

____

____

Name: Date:
Instructor: Section:

Chapter 1 FUNCTION SENSE

Activity 1.5

Learning Objectives
1. Describe in words what a graph tells you about a given situation.
2. Sketch a graph that best represents the situation described in words.
3. Identify increasing, decreasing, and constant parts of a graph.
4. Identify minimum and maximum points on a graph.

Key Terms
Use the vocabulary terms listed below to complete each statement in Exercises 1–2.

maximum **minimum**

1. If a function decreases and then increases, the point where the graph changes from falling to rising is called a __________________ point.

2. If a function increases and then decreases, the point where the graph changes from rising to falling is called a __________________ point.

Practice Exercises
For #3-12, use the following graph that shows an account's return over a 10-year period.

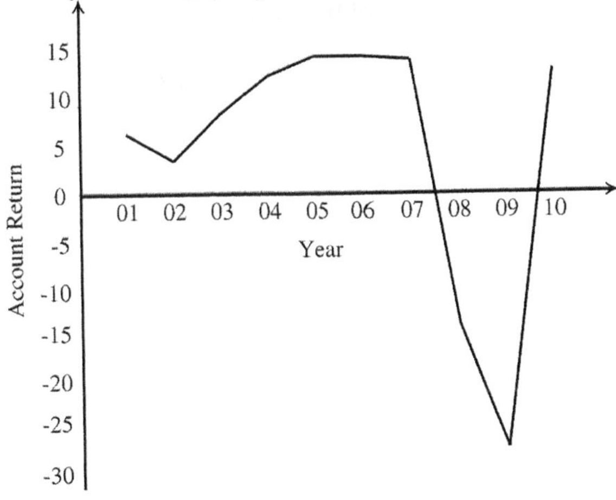

3. Identify the independent variable.

4. Identify the dependent variable.

3. ________________

4. ________________

5. Interpret the situation being represented in the period from 2001 to 2002.

6. Interpret the situation being represented in the period from 2002 to 2005.

7. Interpret the situation being represented in the period from 2005 to 2007.

8. Interpret the situation being represented in the period from 2007 to 2009.

9. Interpret the situation being represented in the period from 2009 to 2010.

10. Interpret the situation being represented in the year 2005.

11. Interpret the situation being represented in the year 2009.

12. Interpret the situation being represented in the period from 2009 to 2010.

5. ______________

____

6. ______________

____

7. ______________

____

8. ______________

____

9. ______________

____

10. ______________

____

11. ______________

____

12. ______________

____

Name: Date:
Instructor: Section:

For #13-16, use the following graph that shows the US car-loan delinquency rate (as a percent) for a time period (in number of months) after origination.

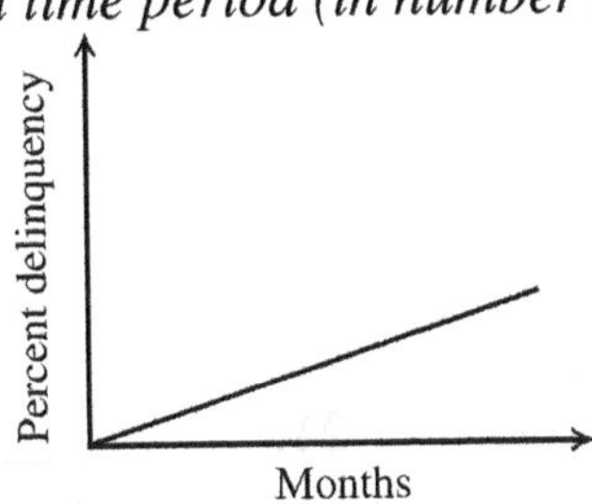

13. Identify the independent variable.

14. Identify the dependent variable.

15. Interpret the situation being represented.

16. Relate your interpretation to the graph.

13. ______________

14. ______________

15. ______________

____

16. ______________

____

For #17-20, use the following graph that shows the percent of the population living on less than $1.25/day from a United Nations report on developing countries.

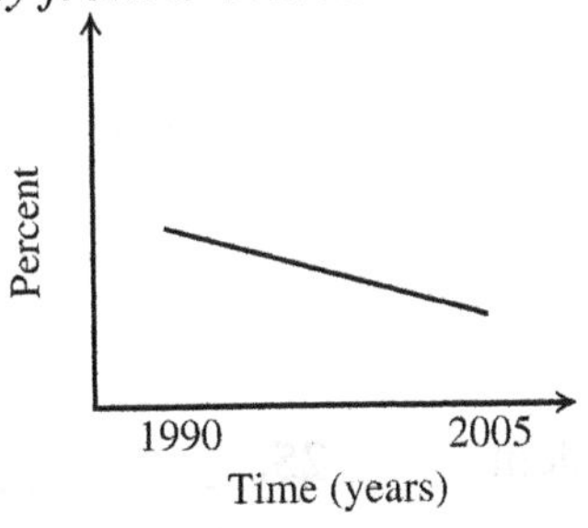

17. Identify the independent variable.

18. Identify the dependent variable.

19. Interpret the situation being represented.

20. Relate your interpretation to the graph.

17. ______________

18. ______________

19. ______________

____

20. ______________

____

For #21-24, use the following graph that shows China's internet use over a 4-year period.

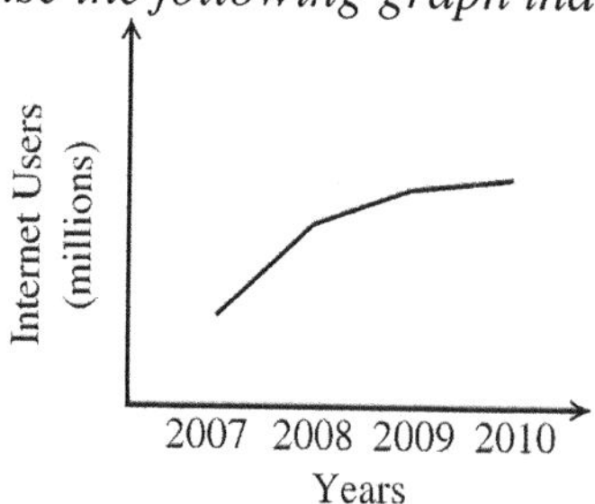

21. Identify the independent variable.

22. Identify the dependent variable.

21. ____________

22. ____________

23. Interpret the situation being represented.

24. Relate your interpretation to the graph.

23. ____________

____

24. ____________

____

For #25-28, use the following graph that shows Great Britain's spending on Defense.

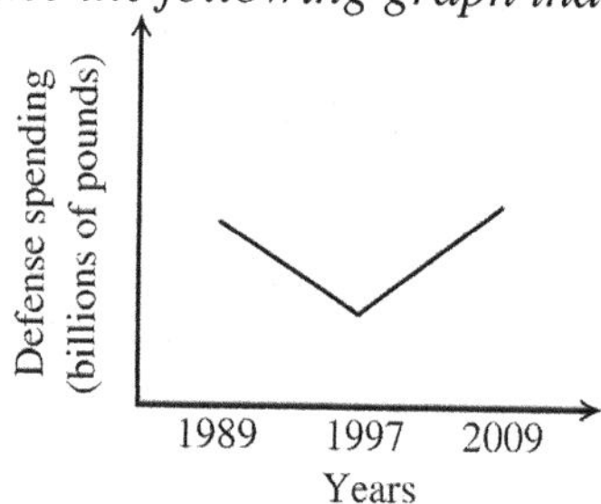

25. Identify the independent variable.

26. Identify the dependent variable.

25. ____________

26. ____________

27. Interpret the situation being represented.

28. Relate your interpretation to the graph.

27. ____________

____

28. ____________

____

Name: Date:
Instructor: Section:

Concept Connections

29. In the graph before Exercise #25, there is a minimum at year 1997. Explain why this is a minimum.

____

____

30. In the graph before Exercise #21, at which year does the maximum occur? Explain why this is a maximum.

____

____

Name: Date:
Instructor: Section:

Chapter 1 FUNCTION SENSE

Activity 1.6

Learning Objectives

1. Determine the average rate of change.

Key Terms

Use the vocabulary terms listed below to complete each statement in Exercises 1–2.

average rate of change **scatterplot**

1. The ________________ indicates how much, and in which direction, the output changes when the input increases by a single unit.

2. A set of points in the plane whose coordinate pairs represent input/output pairs of a data set is called a(n) ________________.

Practice Exercises

For #3-15, the following table of data from the United States Treasury gives the national debt (in trillions of dollars.)

Year	1980	1985	1990	1995	2000	2005	2010
Debt	0.91	1.82	3.23	4.97	5.67	7.93	13.56

3. Determine the average rate of change in the national debt from 2000 to 2010.

4. Describe what the average rate of change in Exercise #3 represents in this situation.

3. ________________

4. ________________

____

____

5. Determine the average rate of change in the national debt from 1985 to 1995.

6. Determine a 5-year period in which the average rate of change is negative.

5. ________________

6. ________________

7. Explain your answer to Exercise #6.

8. During what 5-year period did the average rate of change of the national debt increase the most?

7. ____________________

____

____

8. ____________________

9. In Exercise #8, what is the average rate of change during that 5-year period?

10. In Exercise #9, explain the meaning of the answer.

9. ____________________

10. ____________________

____

____

11. Is the average rate of change zero over any 5-year period?

12. Explain your answer to Exercise #11.

11. ____________________

12. ____________________

____

13. What is the average rate of change over the 30-year period described in the table?

14. Describe what the average rate of change in Exercise #13 represents in this situation.

13. ____________________

14. ____________________

____

____

15. What trend would you observe in the graph during this 30-year period?

15. ____________________

____

Name: Date:
Instructor: Section:

For #16-28, the following table of data from the United States Census Bureau gives the national population (in millions of people).

Year	1910	1920	1930	1940	1950	1960	1970	1980	1990	2000	2010
Population	92	106	123	132	152	181	205	227	249	281	309

16. Determine the average rate of change in the population from 2000 to 2010.

17. Describe what the average rate of change in Exercise #16 represents in this situation.

16. ______________

17. ______________

____

____

18. Determine the average rate of change in the population from 1930 to 1940.

19. Determine a 10-year period in which the average rate of change is negative.

18. ______________

19. ______________

20. Explain your answer to Exercise #19.

21. During what 10-year period did the average rate of change of the national population increase the most?

20. ______________

____

____

21. ______________

22. In Exercise #21, what is the average rate of change during that 10-year period?

23. In Exercise #22, explain the meaning of the answer.

22. ______________

23. ______________

____

____

24. Is the average rate of change zero over any 10-year period?

25. Explain your answer to Exercise #24.

24. ________________

25. ________________

____

26. What is the average rate of change over the 100-year period described in the table?

27. Describe what the average rate of change in Exercise #26 represents in this situation.

26. ________________

27. ________________

____

____

28. What trend would you observe in the graph during this 100-year period?

28. ________________

____

Concept Connections

29. If the rate of change for a certain period is zero, what can be said about the graph in that period?

____

____

30. What do the symbols Δx and Δy represent in the quotient $\frac{\Delta y}{\Delta x}$?

____

____

Name: Date:
Instructor: Section:

Chapter 1 FUNCTION SENSE

Activity 1.7

Learning Objectives
1. Interpret slope as an average rate of change.
2. Use the formula to determine slope.
3. Discover the practical meaning of vertical and horizontal intercepts.
4. Develop the slope-intercept form of an equation of a line.
5. Use the slope-intercept formula to determine vertical and horizontal intercepts.
6. Determine the zeros of a function.

Key Terms
Use the vocabulary terms listed below to complete each statement in Exercises 1–2.

linear function **slope**

1. The constant average rate of change is called the ______________ .

2. A function for which the average rate of change between any pair of points remains constant is called a ______________ .

Practice Exercises
For #3-4, use the function $\{(-3,\ 5),\ (4,\ 2)\ (11,-1)\}$.

3. Is the function linear?

4. What is the average rate of change?

3. ______________

4. ______________

5. Determine whether the following function is linear.

x	3	−5	7
y	2	4	−6

5. ______________

For #6-9, consider the equation $y = -4x + 1$.

6. Construct a table of three ordered pairs that satisfy the equation.

6. ______________

7. What is the slope of the line represented by the equation?

8. What is the vertical intercept?

7. ______________

8. ______________

9. What is the horizontal intercept?

9. ______________

For #10-14, consider the points $(5,-6)$ and $(0,\ 4)$.

10. Determine the vertical intercept of the line.

11. Determine the slope of the line.

10. ______________

11. ______________

12. What is the equation of the line through these points? Use function notation.

13. What is the horizontal intercept?

12. ______________

13. ______________

14. Determine the zeros of the function.

14. ______________

For #15-18, consider the equation $y = 3x - 3$.

15. Identify the slope.

16. What is the vertical intercept?

15. ______________

16. ______________

17. What is the horizontal intercept?

18. Determine the zeros of the function.

17. ______________

18. ______________

Name: Date:
Instructor: Section:

For #19-21, consider the points $(0,-2)$ *and the slope* $m = \frac{1}{2}$.

19. What is the equation of the line?

20. What is the horizontal intercept?

19. ______________

20. ______________

21. Determine the zeros of the function.

21. ______________

For #22-25, consider the points $(-3,\ 4)$ and $(0,\ 1)$.

22. Determine the slope of the line.

23. Find the vertical intercept.

22. ______________

23. ______________

24. What is the equation of the line that goes through the points? Use function notation.

25. What is the horizontal intercept?

24. ______________

25. ______________

For #26-28, use your graphing calculator to graph the linear functions defined by the following equations. Discuss the similarities and differences of the graphs.

26. $y = -\frac{1}{3}x - 5,\ y = -\frac{1}{3}x,\ y = -\frac{1}{3}x + 1$

26. ______________

____

27. $y = 7x + 3,\ y = -4x + 3,\ y = 3$

27. ______________

____

28. $y = 5x,\ y = -\frac{2}{5}x,\ y = -8x$

28. ______________

____

Concept Connections

29. What are the vertical intercept and the horizontal intercept?

____

____

30. Describe the graph of a linear function that has a positive slope. Describe the graph of a linear function that has a negative slope.

____

____

Name: Date:
Instructor: Section:

Chapter 1 FUNCTION SENSE

Activity 1.8

Learning Objectives

1. Write a linear equation in the slope-intercept form, given the initial value and the average rate of change.
2. Write a linear equation given two points, one of which is the vertical intercept.
3. Use the point-slope form to write a linear equation given two points, neither of which is the vertical intercept.
4. Compare slopes of parallel lines.

Key Terms

Use the vocabulary terms listed below to complete each statement in Exercises 1–2.

point-slope **slope-intercept**

1. The equation $y = 3x - 1$ is given in ________________ form.

2. The equation $y = 2 + 3(x - 1)$ is in ________________ form.

Practice Exercises

For #3-28, find the equation of the line. The final equation should be solved for the output y.

3. Slope $= -\frac{2}{3}$
vertical intercept = 13

4. Slope $= -8$
vertical intercept = 7

3. ________________

4. ________________

5. Containing the points (–5, 4) and (0, 8)

6. Containing the points (12, –6) and (0, 3)

5. ________________

6. ________________

7. Slope $= -2$
contains the point (–4, 5)

8. Slope $= 0.5$
contains the point (6, –2)

7. ________________

8. ________________

9. Containing the points (–9, –6) and (–7, 4)

10. Containing the points (10, 6) and (5, –24)

9. ______________

10. ______________

11. Parallel to the line $y = -5x + 3$, passing through the point (4, 7)

12. Parallel to the line $y = 0.7x - 15$, passing through the point (10, 8)

11. ______________

12. ______________

13. Slope $= \frac{7}{3}$ vertical intercept = 3

14. Containing the points (–7, 9) and (0, 5)

13. ______________

14. ______________

15. Slope $= -9$ contains the point (–8, –3)

16. Containing the points (–66, –76) and (–58, 44)

15. ______________

16. ______________

17. Parallel to the line $y = \frac{1}{5}x + 163$, passing through the point (20, –8)

18. Containing the points (–16, –33) and (8, 15)

17. ______________

18. ______________

Name: Date:
Instructor: Section:

19. Containing the points (11, 19) and (0, –25)

20. Slope $= 32$ vertical intercept = 75

19. ______________

20. ______________

21. Containing the points (72, 67) and (0, 58)

22. Parallel to the line $y = -x$, passing through the point (26, –28)

21. ______________

22. ______________

23. Containing the points (–0.7, –0.5) and (–1.7, 0.5)

24. Slope $= 13$ contains the point (–16, –32)

23. ______________

24. ______________

25. Containing the points (1.1, –7.4) and (0, 3.6)

26. Parallel to the line $y = -7x - 1$, passing through the point (1, –7)

25. ______________

26. ______________

27. Slope $= 12$ contains the point (3, 23)

28. Slope $= 300$ vertical intercept = 135

27. ______________

28. ______________

Concept Connections

29. Give the definition of parallel lines.

____

____

30. For a linear function, what can be said about the average rate of change?

____

____

Name: Date:
Instructor: Section:

Chapter 1 FUNCTION SENSE

Activity 1.9

Learning Objectives
1. Write an equation of a line in standard form $Ax + By = C$.
2. Write the slope-intercept form of a linear equation given the standard form.
3. Determine the equation of a horizontal line.
4. Determine the equation of a vertical line.

Key Terms
Use the vocabulary terms listed below to complete each statement in Exercises 1–2.

horizontal **vertical**

1. A ________________ line has a slope that is undefined.

2. A ________________ line has a slope that is zero.

Practice Exercises
For #3-6, consider the linear equation $4x - 5y = 20$.

3. Write the linear equation in slope-intercept form.

4. Determine the slope.

5. Determine the vertical intercept.

6. What is the horizontal intercept?

3. ________________

4. ________________

5. ________________

6. ________________

For #7-10, consider the linear equation $-2x + 3y = 6$.

7. Write the linear equation in slope-intercept form.

8. Determine the slope.

7. ________________

8. ________________

9. Determine the vertical intercept.

10. What is the horizontal intercept?

9. ______________

10. ______________

For #11-14, consider the linear equation $0x-5y=25$.

11. Write the linear equation in slope-intercept form.

12. Determine the slope.

11. ______________

12. ______________

13. Determine the vertical intercept.

14. What is the horizontal intercept?

13. ______________

14. ______________

For #15-20, consider the horizontal line through the point $(4,-6)$.

15. Write the equation of the horizontal line.

16. What is the slope of the line?

15. ______________

16. ______________

17. What is the vertical intercept of the line?

18. What is the horizontal intercept of the line?

17. ______________

18. ______________

19. Does the graph represent a function?

20. Explain your answer to Exercise #19.

19. ______________

20. ______________

____

Name: Date:
Instructor: Section:

For #21-26, consider the vertical line through the point $(8,-1)$.

21. Write the equation of the vertical line.

22. What is the slope of the line?

21. ______________

22. ______________

23. What is the vertical intercept of the line?

24. What is the horizontal intercept of the line?

23. ______________

24. ______________

25. Does the graph represent a function?

26. Explain your answer to Exercise #25.

25. ______________

26. ______________

____

For #27-28, determine three ordered pairs that satisfy each of the following equations.

27. $g(x)=4$

28. $3x+6=0$

27. ______________

28. ______________

Concept Connections

29. Explain standard form of a linear equation.

____

____

30. What is the general form of every horizontal line? What is the general form of every vertical line?

____

____

Name: Date:
Instructor: Section:

Chapter 1 FUNCTION SENSE

Activity 1.10

Learning Objectives

1. Construct scatterplots from sets of data pairs.
2. Recognize when patterns of points in a scatterplot have a linear form.
3. Recognize when the pattern in the scatterplot shows that the two variables are positively related or negatively related.
4. Estimate and draw a line of best fit through a set of points in a scatterplot.
5. Use a graphing calculator to determine a line of best fit by the least-squares method.
6. Measure the strength of the correlation (association) by a correlation coefficient.
7. Recognize that a strong-correlation does not necessarily imply a linear or a cause-and-effect relationship.

Practice Exercises

For #1-9, use the following data points.

x	0	2	4	6	8	10
$f(x)$	−9.4	−1.9	5.6	13.2	20.7	28.2

1. Plot the data points.

2. With a straight edge, draw a line that you think looks like the line of best fit. Does this data appear to be linear?

3. Use your graphing calculator to determine the equation of regression.

2. ______________

3. ______________

4. What is the correlation coefficient?*
(If your calculator does not automatically produce this, see the note below.)

5. What is the slope of the line?

4. ______________

5. ______________

6. Use the equation from Exercise #3 to predict the value of y when $x = 15$.

7. Use the equation from Exercise #3 to predict the value of y when $x = 5$.

6. ______________

7. ______________

8. Which prediction $f(5)$ or $f(15)$ would be more accurate? Explain.

____

____

9. Use the linear model from Exercise #3 to find x when $f(x) = 0$.

9. ______________

---

*If your graphing calculator does not give r, follow these steps.

1. Press 2nd, Catalog (the second function of the number zero)
2. Scroll down the list to DiagnosticOn.
3. Press Enter twice.

At this point your graphing calculator will include r in all future regression models.

To turn the correlation coefficient off, follow the same steps except choosing DiagnosticOff.

Name: Date:
Instructor: Section:

For #10-18, use the following data points.

x	0	3	6	9	12
$f(x)$	16.4	3.8	−8.8	−21.4	−34.2

10. Plot the data points.

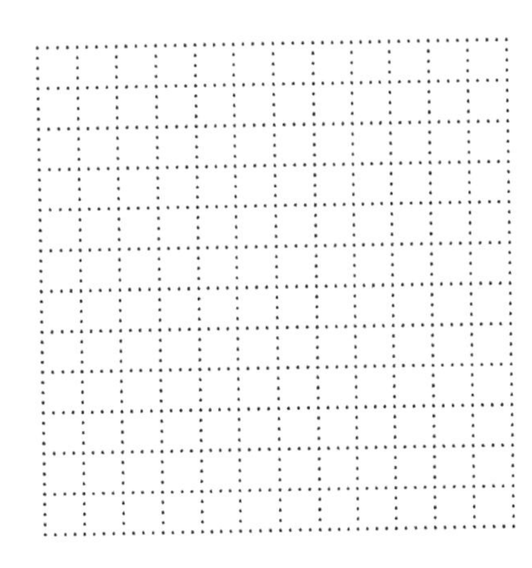

11. With a straight edge, draw a line that you think looks like the line of best fit. Does this data appear to be linear?

12. Use your graphing calculator to determine the equation of regression.

11. ______________

12. ______________

13. What is the correlation coefficient?

14. What is the slope of the line?

13. ______________

14. ______________

15. Use the equation from Exercise #12 to predict the value of y when $x = 5$.

16. Use the equation from Exercise #12 to predict the value of y when $x = -5$.

15. ______________

16. ______________

17. Which prediction, $f(5)$ or $f(-5)$, would be more accurate?

18. Use the linear model from Exercise #12 to find x when $f(x) = 0$.

17. ______________

18. ______________

For #19-28, use the following data points.

x	0	5	10	15	20
$f(x)$	−42.67	76.67	108.33	185.45	260.98

19. Plot the data points.

20. With a straight edge, draw a line that you think looks like the line of best fit. Does this data appear to be linear?

21. Use your graphing calculator to determine the equation of regression.

20. ______________

21. ______________

22. What is the correlation coefficient?

23. What is the slope of the line?

22. ______________

23. ______________

24. Use the equation from Exercise #21 to predict the value of y when $x = 2$.

25. Use the equation from Exercise #21 to predict the value of y when $x = 25$.

24. ______________

25. ______________

26. Which prediction $f(2)$ or $f(25)$ would be more accurate?

27. Use the linear model from Exercise #21 to find x when $f(x) = 0$.

26. ______________

27. ______________

Name: Date:
Instructor: Section:

28. Explain your answer to Exercise #26.

___

___

Concept Connections

29. What is the difference between interpolation and extrapolation?

___

___

___

30. What is a correlation coefficient? What does its value mean?

___

___

___

___

Name: Date:
Instructor: Section:

Chapter 1 FUNCTION SENSE

Activity 1.11

Learning Objectives

1. Solve a system of 2×2 linear equations numerically and graphically.
2. Solve a system of 2×2 linear equations using the substitution method.
3. Solve an equation of the form $ax + b = cx + d$ for x.

Key Terms

Use the vocabulary terms listed below to complete each statement in Exercises 1–3.

consistent **dependent** **inconsistent**

1. A linear system is ________________ if there is no solution.

2. A linear system is ________________ if there is at least one solution.

3. A linear system is ________________ if there are infinitely many solutions.

Practice Exercises

For #4-9, solve the system numerically. Complete the table and state the answer.

4. $y = 2x + 1$
$y = -x + 4$

x	y_1	y_2
-2		
-1		
0		
1		
2		
3		

4. ________________

5. $y = x + 1$
$y = -2x + 10$

x	y_1	y_2
-2		
-1		
0		
1		
2		
3		

5. ________________

6. $y = 2x + 3$

$y = x + 2$

x	y_1	y_2
-2		
-1		
0		
1		
2		
3		

6. ________________

7. $y = -4x + 5$

$y = x + 5$

x	y_1	y_2
-2		
-1		
0		
1		
2		
3		

7. ________________

8. $y = 2x + 6$

$y = 3x + 3$

x	y_1	y_2
-2		
-1		
0		
1		
2		
3		

8. ________________

9. $y = -x + 2$

$y = -x - 3$

x	y_1	y_2
-2		
-1		
0		
1		
2		
3		

9. ________________

Name: Date:
Instructor: Section:

For #10-13, solve the system graphically.

10. $y = 2x - 5$

$y = -x + 1$

10. ______________

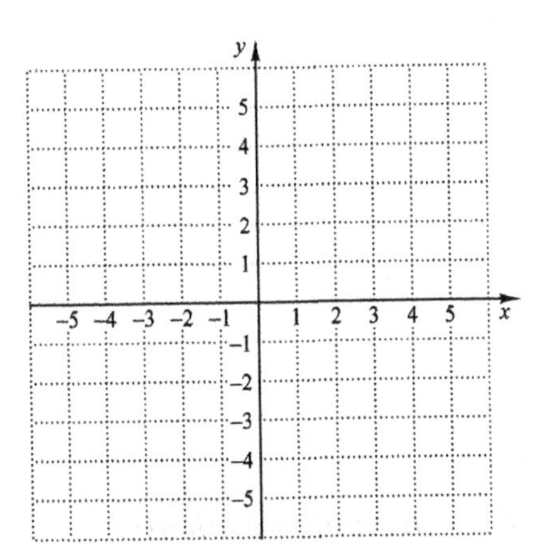

11. $y = -3x + 1$

$y = -x + 3$

11. ______________

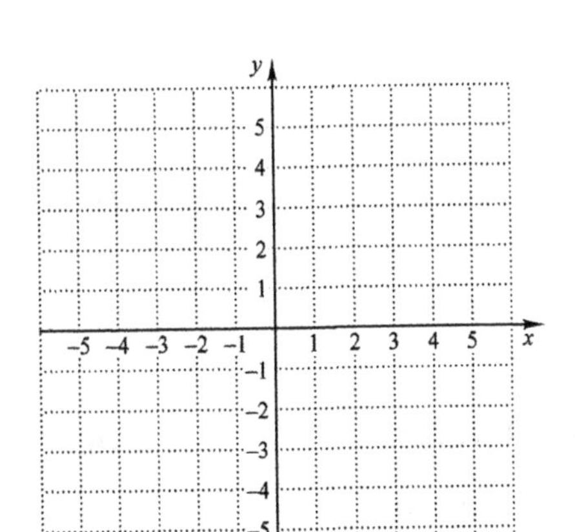

12. $y = 3x + 2$

$y = 3x - 1$

12. ______________

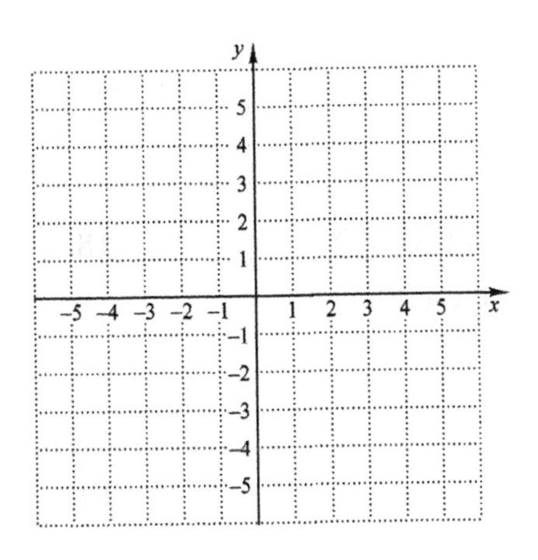

13. $y = x - 2$

$y = -x + 4$

13. 

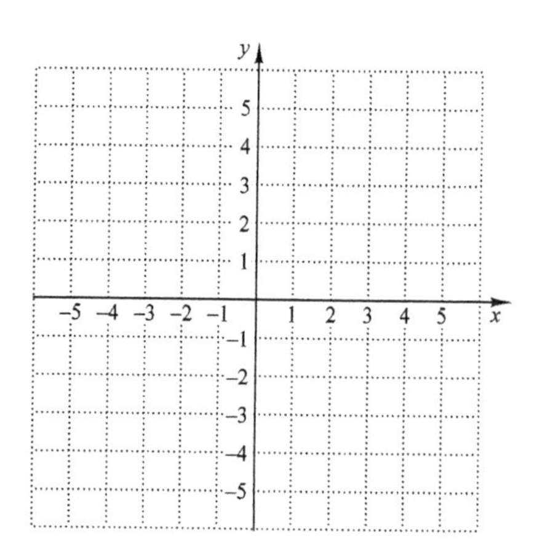

For #14-21, solve the system algebraically (substitution method).

14. $y = 2x + 1$

$y = -x - 2$

15. $y = x - 4$

$y = -2x - 8$

14. ____________________

15. ____________________

16. $y = -4x + 7$

$y = -2x + 1$

17. $y = 3.5x + 6$

$y = 0.5x - 3$

16. ____________________

17. ____________________

18. $y = -2x - 5$

$y = -2x + 7$

19. $y = -1.5x + 2$

$y = 4x + 13$

18. ____________________

19. ____________________

Name: Date:
Instructor: Section:

20. $y = 14.6x - 6.8$
$y = -8.9x + 181.2$

21. $y = -7.1x + 15$
$y = 3.2x - 36.5$

20. ______________

21. ______________

For #22-28, solve each equation.

22. $14x + 7 = 13x + 8$

23. $-4x + 5 = x + 20$

22. ______________

23. ______________

24. $36n - 4 = 35n + 56$

25. $18x - 30 = -12x + 57$

24. ______________

25. ______________

26. $x - 0.5 = -2x + 1$

27. $5x + 25 = 3x + 9$

26. ______________

27. ______________

28. $-4.2x + 20 = 3.5x - 33.9$

28. ______________

Concept Connections

29. Explain graphically what it means for a system of linear equations to be consistent, inconsistent and dependent.

____

____

____

30. When solving a system of linear equations numerically, graphically, or algebraically, which method is most accurate? Why?

____

____

Name: Date:
Instructor: Section:

Chapter 1 FUNCTION SENSE

Activity 1.12

Learning Objectives

1. Solve a 2×2 linear system algebraically using the substitution method and the addition method.
2. Solve equations containing parentheses.

Practice Exercises

For #1-6, solve each of the following equations.

1. $3(x+4)-5x=2x+4+4x$

1. ____________

2. $-4(1-3x)-5(x-2)=-11-4x-5$

2. ____________

3. $2(x-3)+6x=7x+4$

4. $3(2x-1)+4=2x+3$

3. ____________

4. ____________

5. $5(3-4x)+6=x+7$

6. $7-2(x+1)=4x-7$

5. ____________

6. ____________

For #7-17, solve each of the following systems algebraically using the substitution method.

7. $x+y=5$
$y=x+1$

8. $x=y+6$
$x+2y=12$

7. ____________

8. ____________

9. $y = 3x - 2$
$2y - x = 1$

10. $x + y = -6$
$y - x = 8$

9. ______________

10. ______________

11. $x - y = -5$
$2x = y + 1$

12. $x - 3y = 7$
$3x - 5y = 1$

11. ______________

12. ______________

13. $7x + 2y = -4$
$8x - y = 2$

14. $x + y = 5$
$x = 3 + y$

13. ______________

14. ______________

15. $2x - 3y = 5$
$5x + y = 4$

16. $3x + 5y = 3$
$x + 4y = 8$

15. ______________

16. ______________

Name: Date:
Instructor: Section:

17. $x + 10y = 306$
$y + 10x = 90$

17. ______________

For #18-28, solve each of the following systems algebraically using the addition method.

18. $x - y = 3$
$x + y = 11$

19. $x + y = 16$
$-x + 5y = 14$

18. ______________

19. ______________

20. $3x - y = 3$
$-5x + y = 3$

21. $x + 3y = 7$
$-x + 4y = 7$

20. ______________

21. ______________

22. $2x - 3y = 18$
$2x + 3y = -6$

23. $x + y = 9$
$2x - y = -3$

22. ______________

23. ______________

24. $9x+5y=6$
$2x-5y=-17$

25. $0.12x-0.06y=12$
$0.08x+0.16y=-16$

24. ______________

25. ______________

26. $6x-0.5y=9$
$1.5x+2.25y=-12$

27. $x+4y=5$
$-3x+2y=13$

26. ______________

27. ______________

28. $2x+y=-7$
$x+2y=1$

28. ______________

Concept Connections

29. Explain the steps to solve a linear system by substitution.

____

____

30. Explain the steps to solve a linear system by addition.

____

____

Name:
Instructor:

Date:
Section:

Chapter 1 FUNCTION SENSE

Activity 1.13

Learning Objectives

1. Solve a 3×3 linear system of equations.

Key Terms

Use the vocabulary terms listed below to complete each statement in Exercises 1–2.

inconsistent **dependent**

1. A(n) ________________ system of equations has no solution.

2. A(n) ________________ system of equation has infinitely many solutions.

Practice Exercises

For #3-12, solve each system of equations.

3.
$$\begin{aligned} x-y-2z&=1\\ x-5y+2z&=5\\ 2x-3y-4z&=2 \end{aligned}$$

4.
$$\begin{aligned} x+y-z&=4\\ 2x-y-3z&=1\\ x+2y+3z&=-1 \end{aligned}$$

3. ________________

4. ________________

5.
$$\begin{aligned} x-2y-z&=4\\ 2x+3y+2z&=11\\ 2x-y+4z&=5 \end{aligned}$$

6.
$$\begin{aligned} x+2y-3z&=8\\ 2x+y-2z&=11\\ x+5y-8z&=15 \end{aligned}$$

5. ________________

6. ________________

7.
$$\begin{aligned} 2x - y + z &= 5 \\ x - y - z &= 1 \\ 6x + 3y - 2z &= 10 \end{aligned}$$

8.
$$\begin{aligned} x + 6y + 3z &= 4 \\ 2x + y + 2z &= 3 \\ 3x - 2y + z &= 0 \end{aligned}$$

7. ____________________

8. ____________________

9.
$$\begin{aligned} x + 2y + z &= 1 \\ x + 5y + 3z &= 2 \\ 7x + 3y - z &= -2 \end{aligned}$$

10.
$$\begin{aligned} x - y + 4z &= 5 \\ 2x + 3y - z &= -5 \\ 4x + y + 3z &= 5 \end{aligned}$$

9. ____________________

10. ____________________

11.
$$\begin{aligned} x + y + z &= 2 \\ -x + 2y + 2z &= 1 \\ 2x - y + 5z &= -5 \end{aligned}$$

12.
$$\begin{aligned} x + 3y + 8z &= 22 \\ 2x - 3y + z &= 5 \\ 3x - y + 2z &= 12 \end{aligned}$$

11. ____________________

12. ____________________

Name: Date:
Instructor: Section:

For #13-16, solve each system of equations, given one variable.

13. If $x = -2$, solve.

$$2x - y - 4z = -12$$
$$x + 2y + 4z = 10$$
$$2x + y + z = 1$$

14. If $z = 3$, solve.

$$x + y - 3z = -10$$
$$2x + y + z = 6$$
$$3x - 2y - 5z = 7$$

13. ______________

14. ______________

15. If $y = -3$, solve.

$$x + y + z = 0$$
$$-x + 2y - 3z = -1$$
$$2x + 3y + 2z = -3$$

16. If $x = 3$, solve.

$$x + y + z = 2$$
$$6x - 4y + 5z = 31$$
$$5x + 2y + 2z = 13$$

15. ______________

16. ______________

For #17-20, try to solve these linear systems. Identify each system as either dependent or inconsistent.

17.

$$x + 2y - 3z = 5$$
$$2x + 5y + 2z = -1$$
$$5x + 12y + z = 10$$

18.

$$x - 3y + 2z = 4$$
$$3x - 5y + 2z = 4$$
$$5x - 11y + 6z = 12$$

17. ______________

18. ______________

19.
$$\begin{aligned} x + y \quad &= 2 \\ 3x + 5y - z &= 3 \\ -x + 5y - 3z &= 4 \end{aligned}$$

20.
$$\begin{aligned} x - 3y + 2z &= 0 \\ 4x - 11y + 2z &= 0 \\ 2x - 5y - 2z &= 0 \end{aligned}$$

19. ____________________

20. ____________________

For #21-28, determine whether the point is a solution to the system of equations.

21. Is (1, 2, 3) a solution of
$$\begin{aligned} x + 2y - 3z &= -4 \\ -2x + 4y + z &= 9 \\ 5x + 3y - 2z &= 5 \end{aligned}$$

22. Is (–3, 0, 5) a solution of
$$\begin{aligned} 2x + y + z &= -2 \\ 2x - y + 3z &= 6 \\ 3x - 5y + 4z &= 7 \end{aligned}$$

21. ____________________

22. ____________________

23. Is (4, 2, 1) a solution of
$$\begin{aligned} 3x + 2y + z &= 17 \\ -3x + y + 2z &= -8 \\ -2x + 5y + 3z &= 5 \end{aligned}$$

24. Is (1, 4, –6) a solution of
$$\begin{aligned} 2x + y + z &= -1 \\ 4x - y - z &= 4 \\ 6x - 3y - 2z &= 3 \end{aligned}$$

23. ____________________

24. ____________________

Name: Date:
Instructor: Section:

25. Is (2, –5, 6) a solution of

$$x+4y+3z=0$$
$$3x+2y+2z=-8$$
$$2x+y+2z=11$$

26. Is (2, –2, 4) a solution of

$$x+y+z=4$$
$$3x-7y+4z=8$$
$$5x+2y-3z=2$$

25. ______________

26. ______________

27. Is (3, 4, –2) a solution of

$$x+8y-6z=-47$$
$$3x-2y+7z=13$$
$$7x-9y-9z=-3$$

28. Is (–1, –2, 3) a solution of

$$x+2y+3z=4$$
$$-2x+3y+z=-1$$
$$-5x+y-2z=-3$$

27. ______________

28. ______________

Concept Connections

29. Explain what a 3×3 system of linear equations means.

30. Explain what a linear equation in three variables means.

Name: Date:
Instructor: Section:

Chapter 1 FUNCTION SENSE

Activity 1.14

Learning Objectives
1. Solve a linear system of equations using matrices.

Key Terms
Use the vocabulary terms listed below to complete each statement in Exercises 1–4.

augmented matrix	**matrix**
elementary row operations	**reduced row echelon form**

1. ____________________ are interchange two equations, multiply one equation by a nonzero constant, and add a multiple of one row to another equation and replace the second equation.

2. The ____________________ includes all the coefficients with the last column containing the constant terms.

3. A matrix is said to be in ____________________ if the matrix has 1s down the main diagonal and 0s above and below each 1.

4. Any rectangular array of numbers or symbols is called a(n) ____________________.

Practice Exercises
For #5-12, write the augmented matrix for each system of linear equations.

5. $8x+4y=3$
$2x-3y=1$

6. $x-6y=0$
$3x+y=-5$

5. ____________________

6. ____________________

7. $4x-3y-z=1$
$2x+2y+z=5$
$8x-y+z=5$

8. $2x+2y+3z=3$
$x-2y-z=-5$
$x-4y+2z=0$

7. ____________________

8. ____________________

9. $$\begin{aligned} 7x-3y-5z&=14 \\ y+2z&=2 \\ x-3z&=6 \end{aligned}$$

10. $$\begin{aligned} x-y+z&=4 \\ x+4y&=8 \\ x-2z&=-1 \end{aligned}$$

9. ________________

10. ________________

11. $$\begin{aligned} x+7y-z&=8 \\ y+6z&=-2 \\ z&=1 \end{aligned}$$

12. $$\begin{aligned} x-3y-4z&=-1 \\ y-2z&=2 \\ z&=-3 \end{aligned}$$

11. ________________

12. ________________

For #13-18, write the system of linear equations represented by the augmented matrix. Use variables x, y, and z when appropriate.

13. $$\begin{bmatrix} 2 & -8 & 5 \\ 1 & 3 & -1 \end{bmatrix}$$

14. $$\begin{bmatrix} 1 & 1 & -3 \\ 2 & -5 & 0 \end{bmatrix}$$

13. ________________

14. ________________

15. $$\begin{bmatrix} 1 & -4 & 1 & 3 \\ 2 & 1 & -2 & 1 \\ 1 & -2 & -5 & 10 \end{bmatrix}$$

16. $$\begin{bmatrix} 2 & 1 & -2 & 6 \\ 1 & -3 & -3 & 8 \\ 2 & 9 & 2 & -5 \end{bmatrix}$$

15. ________________

16. ________________

Name: Date:
Instructor: Section:

17. $\begin{bmatrix} 1 & 4 & -2 & 0 \\ 0 & 1 & 4 & -5 \\ 1 & 0 & -3 & 0 \end{bmatrix}$

18. $\begin{bmatrix} 1 & -2 & 3 & -4 \\ 0 & 1 & -5 & 6 \\ 0 & 0 & 1 & -8 \end{bmatrix}$

17. ______________

18. ______________

For #19-20, a system of equations has the following row-reduced augmented matrix. Write the solution of the system.

19. $\begin{bmatrix} 1 & 0 & 0 & -4 \\ 0 & 1 & 0 & 3 \\ 0 & 0 & 1 & -2 \end{bmatrix}$

20. $\begin{bmatrix} 1 & 0 & 0 & 1 \\ 0 & 1 & 0 & 0 \\ 0 & 0 & 1 & 3 \end{bmatrix}$

19. ______________

20. ______________

For #21-22, use your graphing calculator to rewrite each augmented matrix in reduced row echelon form.

21. $\begin{bmatrix} 2 & -3 & 5 & 27 \\ 5 & -1 & 4 & 27 \\ 1 & 2 & -1 & -4 \end{bmatrix}$

22. $\begin{bmatrix} 1 & 1 & 1 & 1 \\ 1 & -2 & -3 & 3 \\ 4 & 5 & 6 & 4 \end{bmatrix}$

21. ______________

22. ______________

For #23-28, use your graphing calculator to solve each system of equations.

23. $\begin{aligned} 3x+2y+3z&=7 \\ 2x+9y+6z&=5 \\ x-y+z&=4 \end{aligned}$

24. $\begin{aligned} x-y+2z&=-3 \\ 2x+y+z&=-3 \\ x+2y+3z&=4 \end{aligned}$

23. ______________

24. ______________

25. $x + 3y - 2z = 6$
$2x - y - z = -3$
$x + y + z = 6$

26. $2x + 2y + z = 11$
$3x + 2y + 2z = 8$
$x + 3y + 4z = 0$

25. ________________

26. ________________

27. $x + 2y - 3z = 9$
$2x - y + 2z = -8$
$3x - y - 4z = 3$

28. $5x + 8y = 1$
$3x + 7y = 5$

27. ________________

28. ________________

Concept Connections

For #29-30, use the following scenario.

At work, the gang often ordered fast food from a local hamburger joint. On Monday, Justin paid $28.65 for 6 double cheeseburgers, 5 orders of fries, and 5 large sodas. On Tuesday, Matt paid $43.15 for 10 double cheeseburgers, 7 orders of fries, and 6 large sodas. On Wednesday, Loren paid $22.15 for 4 double cheeseburgers, 5 orders of fries and 4 large sodas.

29. Find the price for each: a double cheeseburger, an order of fries, and a large soda.

____

____

30. Today, Devon must pay for 8 double cheeseburgers, 6 orders of fries and 8 large sodas. If Devon has $40.00, does he have enough money? Explain.

____

____

Name: Date:
Instructor: Section:

Chapter 1 FUNCTION SENSE

Activity 1.15

Learning Objectives

1. Solve linear inequalities in one variable numerically and graphically.
2. Use properties of inequalities to solve linear inequalities in one variable algebraically.
3. Solve compound inequalities algebraically.
4. Use interval notation to represent a set of real numbers described by an inequality.

Practice Exercises

For #1-4, express each inequality in interval notation.

1. $x < 14$

2. $x \geq -5$

3. $-2.4 < x \leq 13$

4. $-100 \leq x \leq 100$

1. ______________

2. ______________

3. ______________

4. ______________

For #5-8, express each interval as an inequality.

5. $[-5, 9)$

6. $(6, \infty)$

7. $(-\infty, 2]$

8. $(-8.2, -4)$

5. ______________

6. ______________

7. ______________

8. ______________

For #9-20, solve each inequality algebraically.

9. $5x > 35$

10. $8x < -48$

11. $4 - 3x \geq 19$

12. $6 - 5x \geq -14$

13. $1 - 2x < -5$

14. $7 - 4x < 15$

15. $x + 8 \leq 4x - 7$

16. $6x - 5 > 2x + 11$

17. $-1 < 2x - 3 < 5$

18. $-8 < \frac{x}{3} - 1 < 4$

19. $4 \leq 3x + 1 \leq 19$

20. $8 \leq 2 - x \leq 21$

9. ______________

10. ______________

11. ______________

12. ______________

13. ______________

14. ______________

15. ______________

16. ______________

17. ______________

18. ______________

19. ______________

20. ______________

Name: Date:
Instructor: Section:

For #21-24, solve each inequality graphically using your graphing calculator. Round to the nearest tenth.

21. $5-7x \geq 6-2x$

22. $14.6-8.2x \geq 3.5-4.1x$

23. $1.9x-2.5 < 8.1x+7$

24. $0.8x+4.1 \geq 14.6-5x$

21. ____________

22. ____________

23. ____________

24. ____________

For #25-28, graph each interval on a number line.

25. $(3, 8)$

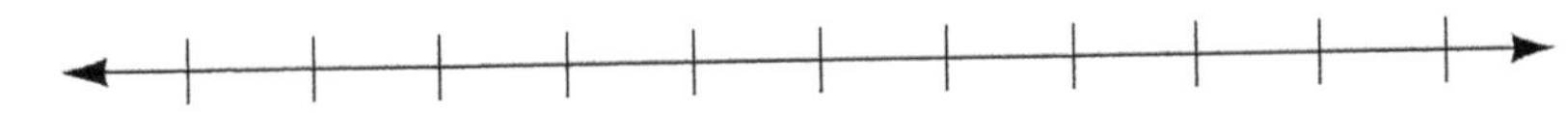

26. $[-2, \infty)$

27. $(-\infty, 1)$

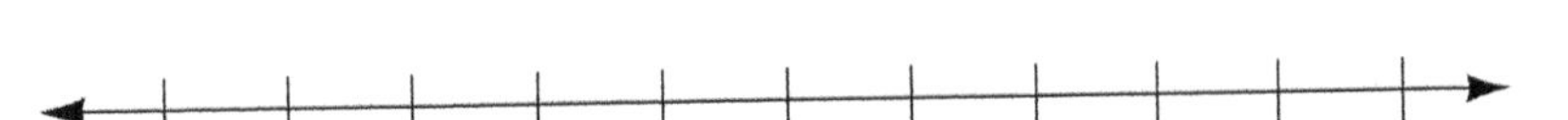

28. $(-4, 5]$

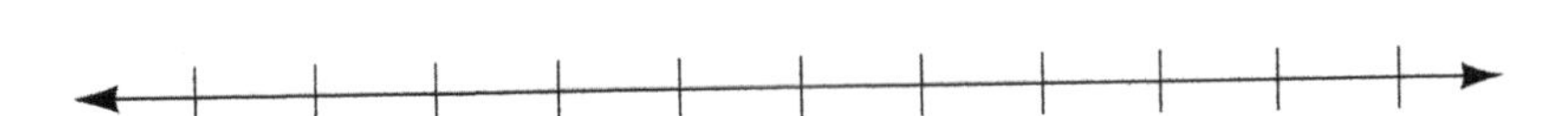

Concept Connections

29. What three approaches are used to solve inequalities?

____

____

30. Explain the difference between open interval, half-open (or half-closed) interval, and a closed interval. Give an example of each.

Name: Date:
Instructor: Section:

Chapter 1 FUNCTION SENSE

Activity 1.16

Learning Objectives
1. Graph a piecewise linear function.
2. Write a piecewise linear function to represent a given situation.
3. Graph a function defined by $y = |x - c|$.

Practice Exercises

For #1-5, use the following piecewise function:

$$f(x) = \begin{cases} -x+2 & \text{if} \quad x < -1 \\ x+3 & \text{if} \quad x \geq -1 \end{cases}$$

1. Sketch the graph of $f(x)$.

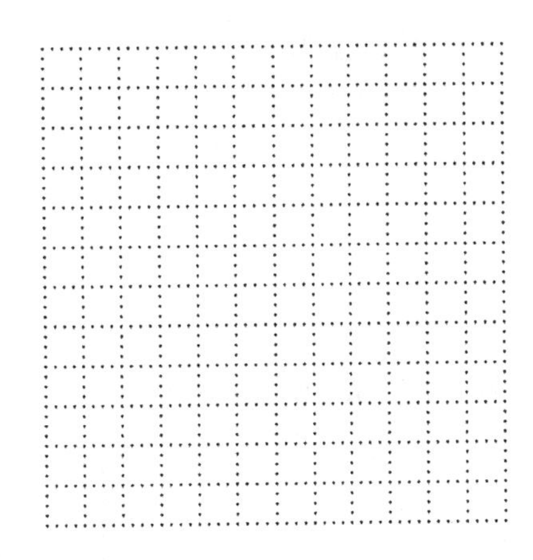

2. What is the domain of the function f?

3. What is the range of the function f?

2. ______________

3. ______________

4. If $f(x) = 6$, find x.

5. If $f(x) = 1$, find x.

4. ______________

5. ______________

For #6-8, use the following piecewise function:

$$g(x)=\begin{cases} x+2 & \text{if} \quad x<0 \\ 1 & \text{if} \quad x\geq 0 \end{cases}$$

6. Sketch the graph of $g(x)$.

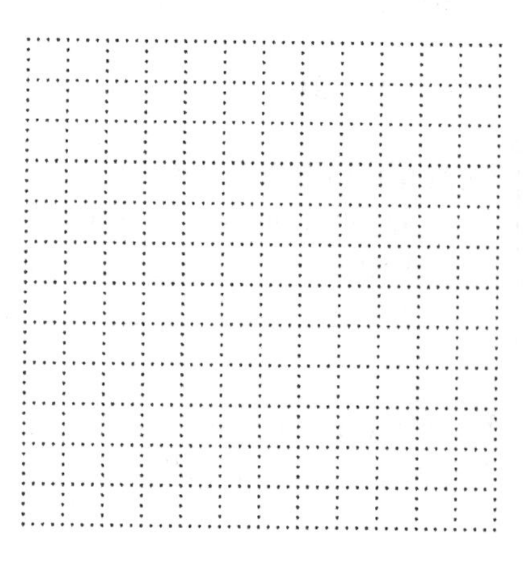

7. What is the domain of the function g ?

8. What is the range of the function g ?

7. ____________

8. ____________

For #9-14, use the following piecewise function:

$$h(x)=\begin{cases} x-2 & \text{if} \quad x<0 \\ -x & \text{if} \quad 0\leq x<3 \\ -x+3 & \text{if} \quad x\geq 3 \end{cases}$$

9. What is the domain of the function h?

10. Determine $h(-1)$.

9. ____________

10. ____________

11. Determine $h(1)$.

12. Determine $h(4)$.

11. ____________

12. ____________

13. Determine $h(0)$.

14. Determine $h(3)$.

13. ____________

14. ____________

Name: Date:
Instructor: Section:

For #15-21, use the following piecewise function:

$$f(x)=\begin{cases}-x-1 & \text{if} \quad x<-1 \\ x+1 & \text{if} \quad x\geq -1\end{cases}$$

15. What is the domain of the function f ?

15. ______________

16. Complete the following table.

x	-3	-2	-1	0	1	2	3
$f(x)$							

17. Sketch the graph of $f(x)$.

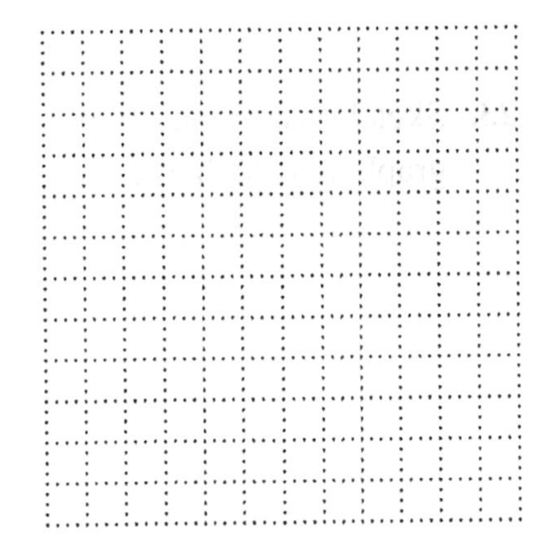

18. Describe the shape of the graph.

19. What is the range of the graph?

18. ______________

19. ______________

20. Sketch that graph of $g(x)=|x+1|$. Verify using your graphing calculator.

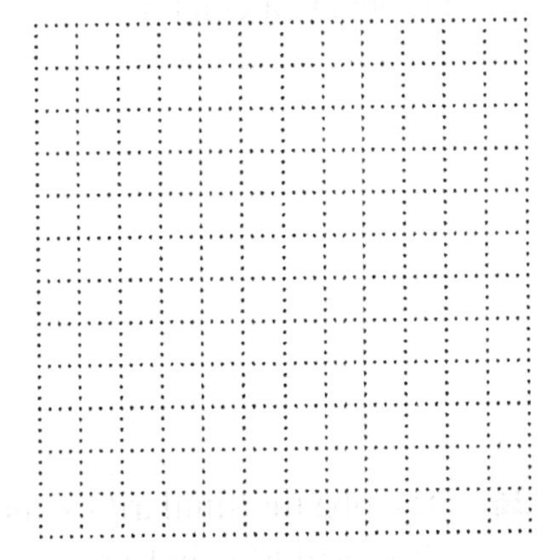

21. Describe the similarities and differences in the two graphs in Exercises #17 and #20.

21. ________________

22. Sketch that graph of $f(x) = |x|$. Verify using your graphing calculator.

23. Sketch that graph of $g(x) = |x| - 4$. Verify using your graphing calculator.

24. Describe the similarities and differences in the two graphs $f(x)$ and $g(x)$ in Exercises #22 and #23.

24. ________________

25. Sketch that graph of $h(x) = |x + 4|$. Verify using your graphing calculator.

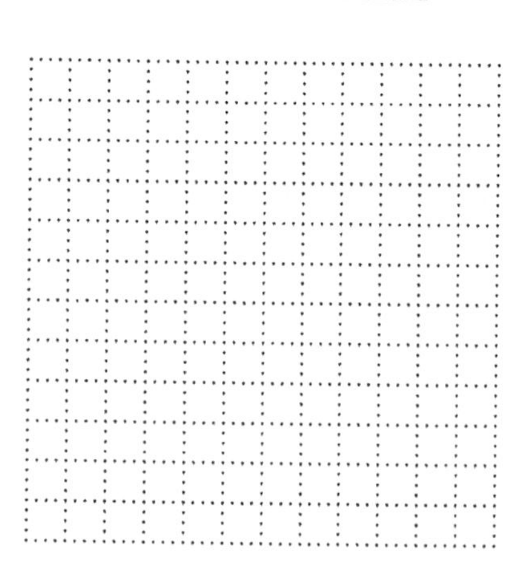

26. Describe the similarities and differences in the two graphs $f(x)$ and $h(x)$ in Exercises #22 and #25.

26. ________________

____

Name:
Instructor:

Date:
Section:

27. Sketch that graph of $k(x) = |x - 3| + 5$. Verify using your graphing calculator.

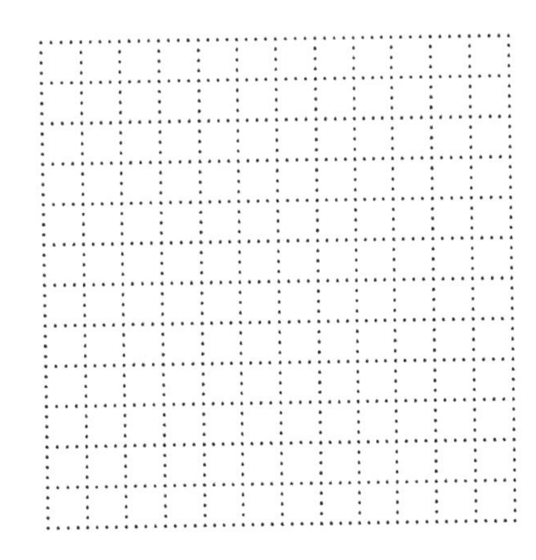

28. Describe the similarities and differences in the two graphs $f(x)$ and $k(x)$ in Exercises #22 and #27.

28. ______________

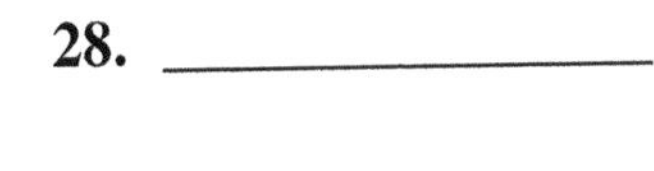

Concept Connections

29. What is a piecewise function?

____

____

30. Describe the graph of an absolute value function $f(x) = |x - b|$.

____

____

Name: Date:
Instructor: Section:

Chapter 2 THE ALGEBRA OF FUNCTIONS

Activity 2.1

Learning Objectives
1. Identify a polynomial expression.
2. Identify a polynomial function.
3. Add and subtract polynomial expressions.
4. Add and subtract polynomial functions.

Key Terms

Use the vocabulary terms listed below to complete each statement in Exercises 1–4.

binomial **polynomial** **trinomial** **monomial**

1. A ________________ is a polynomial with three terms.

2. A ________________ is a polynomial with one term.

3. A ________________ is a polynomial with two terms.

4. Any expression that is formed by adding terms of the form ax^n where a is a real number and n is a nonnegative integer, is called a ________________ expression.

Practice Exercises

For #5-10, write "yes" if the expression is a polynomial in x. Write "no" if the expression is not a polynomial. If the expression is a polynomial, classify it as a monomial, binomial, or trinomial.

5. $4x^{1/2}+3$

6. $7x^8-9x^2$

7. 8

8. $2x^{-2}+6x$

9. $x^2+\frac{1}{3}x+2$

10. $\frac{4}{x}-5$

5. ____________

6. ____________

7. ____________

8. ____________

9. ____________

10. ____________

For #11, suppose that f and g are defined by the following tables.

x	-1	1	3	5	7	9
$f(x)$	-6	-3	0	3	6	9

x	-1	1	3	5	7	9
$g(x)$	-2	0	4	6	8	10

11. Complete the following table.

x	-1	1	3	5	7	9
$f(x)+g(x)$						
$f(x)-g(x)$						

For #12-19, simplify each of the following expressions.

12. $(3x+2)+(5x-7)$

13. $(3x^2-5x+2)-(2x^2-4x-7)$

12. ______________

13. ______________

14. $4(x-3)-5(x+2)$

15. $6x-2-(x^2+x+1)$

14. ______________

15. ______________

16. $3x^2+8x-2(2-x^2)$

17. $(6x-1)-4(2x^2-3x+1)$

16. ______________

17. ______________

18. $3x+4-[2x-3(4-x)]$

19. $5x+3[5x-3(2+6x)]+7$

18. ______________

19. ______________

Name: Date:
Instructor: Section:

For #20-23, given $f(x)=2x-7$ *and* $g(x)=-2x^2+3x-7$, *determine a formula, in simplest form for each of the following.*

20. $f(x)+g(x)$

21. $f(x)-g(x)$

22. $2f(x)+3g(x)$

23. $f(x)-2g(x)$

20. ____________

21. ____________

22. ____________

23. ____________

For #24-28, given $f(x)=x^2-1$ *and* $g(x)=x+3$, *determine a value for each of the following.*

24. $f(2)+g(2)$

25. $g(3)-f(3)$

26. $f(-5)+g(-5)$

27. $f(-2)-g(-2)$

28. $g(0)-f(0)$

24. ____________

25. ____________

26. ____________

27. ____________

28. ____________

Concept Connections

29. Bill says that $\sqrt{3}+x$ and $\sqrt{3+x}$ are both polynomials. Paul says only one is a polynomial. Who is right, and why?

____

____

30. The function for profit is $P(x)=R(x)-C(x)$, where x is the number of items made. The Major Widget company produces 3500 mini-widgets with a profit of $80,500. Write the function using these values.

____

____

Name: Date:
Instructor: Section:

Chapter 2 THE ALGEBRA OF FUNCTIONS

Activity 2.2

Learning Objectives

1. Multiply two binomials using the FOIL method.
2. Multiply two polynomial functions.
3. Apply the property of exponents to multiply powers having the same base.

Practice Exercises

For #1-8, use the properties of exponents to determine each of the following products.

1. $5^3 \cdot 5^8$

2. $y^6 \cdot y$

3. $u^3 v^6$

4. $(3a^5)(5a^3)$

5. $(4x)(-3x^7)(-2x^5)$

6. $(x^2 y^5)(x^3 y^2)$

7. $x^{3n} \cdot x^n$

8. $a^{2x} \cdot a^{3x}$

1. ______________
2. ______________
3. ______________
4. ______________
5. ______________
6. ______________
7. ______________
8. ______________

For #9-16, Determine each product, and simplify the result.

9. $(x+2)(x+8)$

10. $(w-6)(w-4)$

9. ______________

10. ______________

11. $(x-5)(x+8)$

12. $(6+5c)(3+c)$

11. ______________

12. ______________

13. $(x+4)(x-4)$

14. $(7x-1)(3x+4)$

13. ______________

14. ______________

15. $(4x-3)(x-4)$

16. $(x+6)(x-5)$

15. ______________

16. ______________

For #17-20, multiply each expression. Write your answer in simplest form.

17. $(x+3)(x^2-4x+6)$

18. $(x-5)(4x^2-x+3)$

17. ______________

18. ______________

Name: Date:
Instructor: Section:

19. $(x-4)(x^2+4x+16)$

20. $(x+5)(x^2-5x+25)$

19. ______________

20. ______________

For #21-28, expand and simplify the following binomial products.

21. $(x-5)^2$

22. $(x+6)^2$

21. ______________

22. ______________

23. $(5x-4)(5x+4)$

24. $(6x+7)(6x-7)$

23. ______________

24. ______________

25. $(x-25)^2$

26. $(x+13)^2$

25. ______________

26. ______________

27. $(x-11)^2$

28. $(3x-8)(3x+8)$

27. ______________

28. ______________

Concept Connections

29. Identify the Exercises in this Activity that are difference of squares and the Exercises that are a square of a binomial.

____

____

30. FOIL is a term commonly used when multiplying two binomials.
What does FOIL stand for?

____

____

Name: Date:
Instructor: Section:

Chapter 2 THE ALGEBRA OF FUNCTIONS

Activity 2.3

Learning Objectives

1. Convert scientific notation to decimal notation.
2. Convert decimal notation to scientific notation.
3. Apply the property of exponents to divide powers having the same base.
4. Apply the definition of exponents $a^0 = 1$, where $a \neq 0$.
5. Apply the definition of exponents $a^{-n} = \frac{1}{a^n}$, where $a \neq 0$ and n is any real number.

Key Terms

Use the vocabulary terms listed below to complete each statement in Exercises 1–2.

exponential notation **scientific notation**

1. The expression 4^5 is written in ____________________.

2. The expression 1.3×10^8 is written in ____________________.

Practice Exercises

For #3-8, convert each number to scientific notation.

3. 960,000,000,000

4. 0.00000000019

5. 770,000,000,000,000

6. 0.0000000067

7. 10,400,000

8. 0.00038

3. ________________

4. ________________

5. ________________

6. ________________

7. ________________

8. ________________

For #9-14, convert each number to decimal notation.

9. 4.21×10^{-7}

10. 2.03×10^{11}

9. ____________

10. ____________

11. 9×10^{5}

12. 6.06×10^{-10}

11. ____________

12. ____________

13. 7.2×10^{-4}

14. 5.8×10^{8}

13. ____________

14. ____________

For #15-16, divide and write the answer in scientific notation.

15. $\dfrac{9.2\times10^{17}}{2.3\times10^{3}}$

16. $\dfrac{1.26\times10^{17}}{1.5\times10^{4}}$

15. ____________

16. ____________

For #17-28, use the properties of exponents to simplify each of the following. Write your results with positive exponents only.

17. $\dfrac{u^5}{u^2}$

18. $\dfrac{5^8}{5^3}$

17. ____________

18. ____________

19. $\left(\dfrac{2}{b}\right)^0$

20. $\left(\dfrac{1}{w}\right)^{-3}$

19. ____________

20. ____________

Name: Date:
Instructor: Section:

21. $7x^{-4}$

22. $(2x)^{-4}$

21. ________________

22. ________________

23. $\dfrac{12w^8}{4w^4}$

24. $\dfrac{25x^2}{5x^{11}}$

23. ________________

24. ________________

25. $\dfrac{18x^3y^9z^5}{3x^6y^2z^4}$

26. $\dfrac{8a^{-3}b^2}{-2a^{-2}b^0}$

25. ________________

26. ________________

27. $\dfrac{24c^0d^{-4}}{4c^{-5}d}$

28. $x^{-4}(3x^5)(-2x^0)$

27. ________________

28. ________________

Concept Connections

29. Explain the difference between $4x^0$ and $(4x)^0$.

____

____

30. Tony writes the number –17,000,000,000,000 incorrectly in scientific notation as 17×10^{-12}. What is the correct answer?

Name: Date:
Instructor: Section:

Chapter 2 THE ALGEBRA OF FUNCTIONS

Activity 2.4

Learning Objectives

1. Apply the property of exponents to simplify an expression involving a power to a power.
2. Apply the property of exponents to expand the power of a product.
3. Determine the *n*th root of a real number.
4. Write a radical as a power having a rational exponent and write a base to a rational exponent as a radical.

Key Terms

Use the vocabulary terms listed below to complete each statement in Exercises 1–4.

cube root **index** **principal square root** **radicand**

1. In $\sqrt[n]{a} = a^{1/n}$, the number a is called the ________________ .

2. In $\sqrt[n]{a} = a^{1/n}$, the number n is called the ________________ .

3. If a represents a nonnegative real number, the ________________ of a, denoted by $\sqrt{a}$, the number a is defined as the nonnegative number that when squared produces a.

4. The ________________ of any real number a, denoted by $\sqrt[3]{a}$, the number a is defined as the number that when cubed gives a.

Practice Exercises

For #5-12, simplify each expression by applying the properties of exponents. Write your results with positive exponents only.

5. $(x^8)^4$

6. $(2x^2)^4$

5. ________________

6. ________________

7. $(-5x^4)^3$

8. $(x^5)^{-4}$

7. ________________

8. ________________

9. $(-6w^{-7})^2$

10. $(x^{-3}y^4z^{-2})^{-4}$

9. ______________

10. ______________

11. $-(a^8)^5$

12. $(-b^5)^6$

11. ______________

12. ______________

For #13-17, compute each of the following quantities.

13. $64^{1/2}$

14. $36^{1/2}$

13. ______________

14. ______________

15. $(-32)^{3/5}$

16. $81^{-1/2}$

15. ______________

16. ______________

17. $125^{-2/3}$

17. ______________

For #18-20, simplify each of the following.

18. $x^{1/4} \cdot x^{2/3}$

19. $(x^{-1/5})^{-1/3}$

18. ______________

19. ______________

20. $\dfrac{x^{2/3}}{x^{2/5}}$

20. ______________

Name: Date:
Instructor: Section:

For #21-24, write each of the following using fractional exponents.

21. $\sqrt{a}$

22. $\sqrt[6]{w^5}$

21. ______________

22. ______________

23. $\sqrt[3]{a-b}$

24. $\sqrt[7]{xy^5z^2}$

23. ______________

24. ______________

For #25-26, if $f(x)=\sqrt{x-5}$, *determine each of the following.*

25. $f(5)$

26. $f(14)$

25. ______________

26. ______________

For #27-28, if $g(x)=\sqrt[3]{x+6}$, *determine each of the following.*

27. $g(-14)$

28. $g(-6)$

27. ______________

28. ______________

Concept Connections

29. Jeff has a hard time remembering the rules for exponents. How would you help him remember how to simplify the expressions $(x^3)^4$ and y^5y^3?

____

____

30. Name the domain of each of the following functions: $f(x) = \sqrt[3]{x}$ and $g(x) = \sqrt{x}$.

____

____

Name: Date:
Instructor: Section:

Chapter 2 THE ALGEBRA OF FUNCTIONS

Activity 2.5

Learning Objectives

1. Determine the composition of two functions.
2. Explore the relationship between $f(g(x))$ and $g(f(x))$.

Practice Exercises

For #1-5, if $f(x) = 3x + 2$, *determine each of the following.*

1. $f(2)$

2. $f(a)$

3. $f(\text{BOB})$

4. $f(\#)$

5. $f(g(x))$

1. ______________
2. ______________
3. ______________
4. ______________
5. ______________

For #6-9, if $g(x) = 2x^2 - 3x + 1$, *determine each of the following.*

6. $g(3)$

7. $g(a)$

8. $g(\text{BOB})$

9. $g(h(x))$

6. ______________
7. ______________
8. ______________
9. ______________

For #10-14, if $f(x)=4x+1$ and $g(x)=3x-2$, *determine each of the following.*

10. $f(g(x))$ **11.** $f(g(2))$

10. ______________

11. ______________

12. $g(f(x))$ **13.** $g(f(2))$

12. ______________

13. ______________

14. Does $f(g(x))=g(f(x))$?

14. ______________

For #15-19, if $f(x)=5x-2$ and $g(x)=2x-5$, *determine each of the following.*

15. $f(g(x))$ **16.** $f(g(-2))$

15. ______________

16. ______________

17. $g(f(x))$ **18.** $g(f(-2))$

17. ______________

18. ______________

19. Does $f(g(x))=g(f(x))$?

19. ______________

Name: Date:
Instructor: Section:

For #20-24, if $f(x) = -6x + 5$ and $g(x) = 6x + 5$, *determine each of the following.*

20. $f(g(x))$

21. $f(g(-1))$

20. ______________

21. ______________

22. $g(f(x))$

23. $g(f(-1))$

22. ______________

23. ______________

24. Does $f(g(x)) = g(f(x))$?

24. ______________

For #25-28, if $f(x) = x + 3$ and $g(x) = x - 3$, *determine each of the following.*

25. $f(g(x))$

26. $f(g(3))$

25. ______________

26. ______________

27. $g(f(x))$

28. $g(f(3))$

27. ______________

28. ______________

Concept Connections

29. Amy thinks that the order of composition of functions does not matter; in other words $f(g(x))$ is the same as $g(f(x))$. How can you change her mind?

____

____

30. In the situation where $f(g(x)) = g(f(x)) = x$, we say that the functions $f(x)$ and $g(x)$ are inverse functions. Give an example of $f(x)$ and $g(x)$ that fit this situation.

____

____

Name: Date:
Instructor: Section:

Chapter 2 THE ALGEBRA OF FUNCTIONS

Activity 2.6

Learning Objectives

1. Solve problems using the composition of functions.

Practice Exercises

For #1-12, use the following scenario.

The sales tax for a dinner served in a restaurant in Hershey is 6%. Let x represent the price of the dinner. Let $f(x)$ represent the sales tax for dinner.

1. Determine a rule for $f(x)$.

2. Using the rule from Exercise #1, find $f(20.99)$.

3. Interpret the answer from Exercise #2.

4. Let $g(x)$ represent the total of the dinner and the sales tax. Determine a rule for $g(x)$.

5. Using the rule from Exercise #4, find $g(20.99)$.

6. Interpret the answer from Exercise #5.

7. The customary tip is 18% of the dinner. Let $h(x)$ represent the amount of the tip. Determine a rule for $h(x)$.

8. Using the rule from Exercise #7, find $h(20.99)$.

1. ______________

2. ______________

3. ______________

____

4. ______________

5. ______________

6. ______________

____

7. ______________

8. ______________

9. Interpret the answer from Exercise #8.

10. Let $k(x)$ represent the total dinner, sales tax, and tip. Determine a rule for $k(x)$.

9. ________________

____

10. ________________

11. Using the rule from Exercise #10, find $k(20.99)$.

12. Interpret the answer from Exercise #11.

11. ________________

12. ________________

____

For #13-27, use the following scenario.

Skylex Flowers has casual workers deliver flowers on Valentine's Day. Each driver can take 15 floral arrangements per run. Let x represent the total number of arrangements to be delivered on this holiday. Let $f(x)$ represent the number of runs needed.

13. Determine a rule for $f(x)$.

14. Using the rule from Exercise #13, find $f(450)$.

13. ________________

14. ________________

15. Interpret the answer from Exercise #14.

16. Each driver is available to complete 3 runs on Valentine's Day. Let $g(x)$ represent the number of drivers needed for the day. Determine a rule for $g(x)$.

15. ________________

____

16. ________________

17. Using the rule from Exercise #16, find $g(450)$.

18. Interpret the answer from Exercise #17.

17. ________________

18. ________________

____

Name:
Instructor:

Date:
Section:

19. Each driver earns $3.25 for each completed delivery. Let $P(x)$ represent the total pay for deliveries. Determine a rule for $P(x)$.

20. Using the rule from Exercise #19, find $P(450)$.

19. ____________

20. ____________

21. Interpret the answer from Exercise #20.

22. A run is planned so that each delivery is approximately 3 miles apart. Let $d(x)$ represent the total delivery distance. Determine a rule for $d(x)$.

21. ____________

____

22. ____________

23. Using the rule from Exercise #22, find $d(450)$.

24. Interpret the answer from Exercise #23.

23. ____________

24. ____________

____

25. The IRS allows 51 cents per mile for business travel. Let $B(x)$ represent the total mileage expense. Determine a rule for $B(x)$.

26. Using the rule from Exercise #25, find $B(450)$.

25. ____________

26. ____________

27. Interpret the answer from Exercise #26.

27. ____________

____

28. A $150 leather purse is on sale for 25% off. You present a coupon for an additional 15% off. Determine the price you pay for the purse.

28. ______________

Concept Connections

29. Which is a better deal: 15% discount and additional 40% discount, or 20% discount and additional 30% discount?

____

____

30. Which is a better deal: 30% discount or 10% discount and additional 20% discount?

____

____

Name: Date:
Instructor: Section:

Chapter 2 THE ALGEBRA OF FUNCTIONS

Activity 2.7

Learning Objectives

1. Determine the inverse of a function represented by a table of values.
2. Use the notation f^{-1} to represent an inverse function.
3. Use the property $f\left(f^{-1}(x)\right)=f^{-1}\left(f(x)\right)=x$ to recognize inverse functions.
4. Determine the domain and range of a function and its inverse.

Practice Exercises

For #1-13, the function k is defined by the following set of ordered pairs.

{(0, 15), (1, 12), (2, –5), (3, 7), (9, 6)}

1. Write k^{-1} as a set of ordered pairs.

2. Determine the domain of k.

3. Determine the range of k.

4. Determine the domain of k^{-1}.

5. Determine the range of k^{-1}.

6. Determine $k(1)$.

7. Determine $k^{-1}(12)$.

8. Determine $k^{-1}(7)$.

1. ______________

2. ______________

3. ______________

4. ______________

5. ______________

6. ______________

7. ______________

8. ______________

9. Determine $k(3)$.

10. Determine $k\left(k^{-1}(12)\right)$.

9. ______________

10. ______________

11. Determine $k^{-1}\left(k(3)\right)$.

12. Determine $k\left(k^{-1}(x)\right)$.

11. ______________

12. ______________

13. Determine $k^{-1}\left(k(x)\right)$.

13. ______________

For #14-28, use the following scenario.

As a resident of New York City, you are interested in sharing a car for an occasion. One company gives you the information listed in the following table.

Hours rented	1	2	3	4	5	6	7
Cost	30	35	40	45	50	55	60

The cost C is a function of the number of hours the car is rented.

14. Determine $C(3)$.

15. Explain the meaning of Exercise #14 in practical terms.

14.

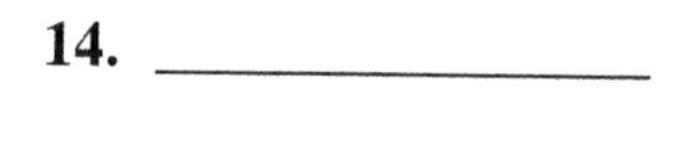

15. ______________

____

16. Determine the domain and range of function C.

16.

Name: Date:
Instructor: Section:

17. You are working out a budget. Use the information from the table to construct another table in which the cost is the input and the number of hours rented is the output.

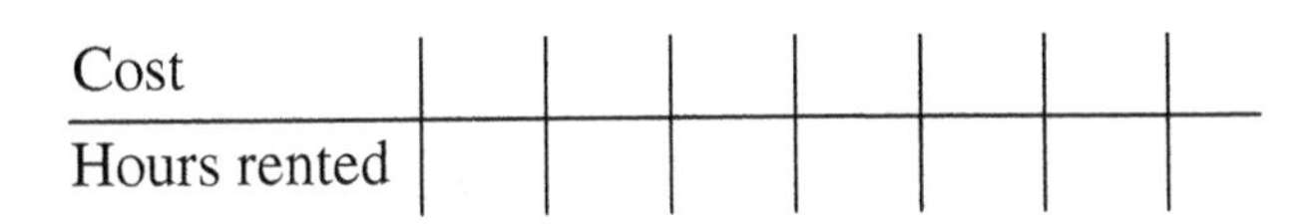

Cost							
Hours rented							

18. Using the input/output values in Exercise #17, determine if the number of hours rented is a function of the cost.

19. From Exercise #17, the number of hours H is a function of the cost. Determine $H(55)$.

18. ______________

19. ______________

20. Explain the meaning of Exercise #19 in practical terms.

21. Determine the domain and range of function H.

20. ______________

____

22. ______________

____

22. How are the domain and range of function C related to the domain and range of function H?

23. Determine $C(7)$.

22. ______________

____

23. ______________

24. Determine $H(60)$.

25. Determine $H(C(7))$.

24. ______________

25. ______________

26. Determine $C(H(30))$. **27.** Determine $H(C(x))$.

26. ______________

27. ______________

28. Determine $C(H(x))$.

28. ______________

Concept Connections

29. What is the difference between $f^{-1}(x)$ and $\frac{1}{f(x)}$?

____

____

30. How can you check that two functions are inverses?

____

____

Name: Date:
Instructor: Section:

Chapter 2 THE ALGEBRA OF FUNCTIONS

Activity 2.8

Learning Objectives

1. Determine the equation of the inverse of a function represented by an equation.
2. Describe the relationship between the graphs of inverse functions.
3. Determine the graph of the inverse of a function represented by a graph.
4. Use the graphing calculator to produce graphs of an inverse function.

Practice Exercises

For #1-16, determine the equation of the inverse of the given function.

1. $f(x) = 3x - 7$

2. $g(x) = 2x - 8$

3. $f(a) = 5 - 6a$

4. $g(z) = 11 - 4z$

5. $h(x) = \frac{x-1}{2}$

6. $f(x) = \frac{x+6}{9}$

7. $g(x) = \frac{4-x}{3}$

8. $h(x) = \frac{8+x}{5}$

1. ________________

2. ________________

3. ________________

4. ________________

5. ________________

6. ________________

7. ________________

8. ________________

9. $w(x)=x+5$

10. $k(x)=12-x$

9. ______________

10. ______________

11. $f(x)=9x+9$

12. $g(x)=7x-7$

11. ______________

12. ______________

13. $h(x)=0.5x+1$

14. $k(x)=0.4x-2$

13. ______________

14. ______________

15. $g(x)=2.5x-5$

16. $f(x)=0.1x+6$

15. ______________

16. ______________

For #17-24, determine whether the given functions are inverses.

17. $f(x)=x+8$ and $g(x)=x-8$

18. $f(x)=2x-7$ and $g(x)=\dfrac{x-7}{2}$

17. ______________

18. ______________

Name: Date:
Instructor: Section:

19. $f(x)=0.4x-2$ and $g(x)=0.4x+0.8$

20. $f(x)=9x+9$ and $g(x)=9x-9$

21. $f(x)=12-x$ and $g(x)=x-12$

22. $f(x)=2x-8$ and $g(x)=0.5x+4$

23. $f(x)=12-6x$ and $g(x)=x+2$

24. $f(x)=x-5$ and $g(x)=x+5$

For #25-26, f and g are inverse functions.

25. If $f(3)=5$, find $g(5)$.

26. If $f(-6)=2$, find $g(2)$.

19. ______________

20. ______________

21. ______________

22. ______________

23. ______________

24. ______________

25. ______________

26. ______________

For #27-28, given the graph of the function, draw the graph of the inverse.

27.

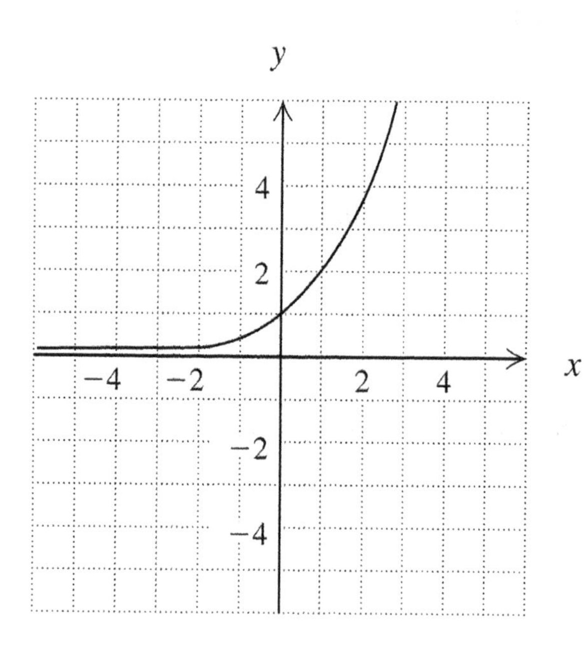

28.

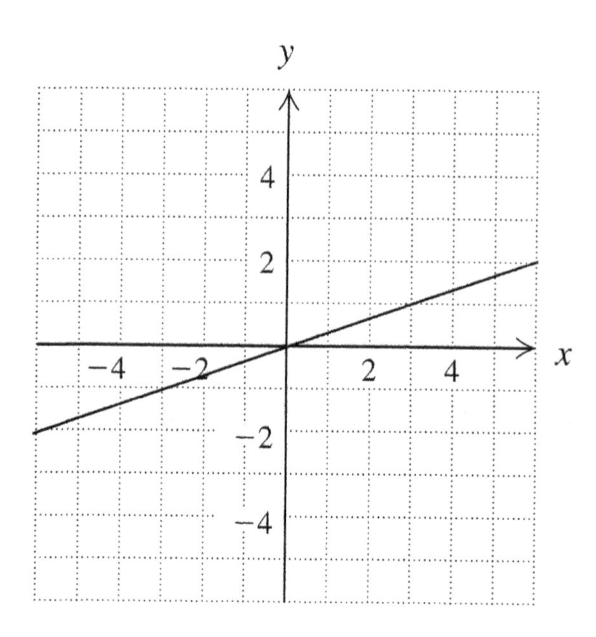

Concept Connections

29. What is the graphical relationship between a function and its inverse?

---

30. What is the relationship between the slopes of a linear function and its inverse?

Name: Date:
Instructor: Section:

Chapter 3 EXPONENTIAL AND LOGORITHMIC FUNCTIONS

Activity 3.1

Learning Objectives

1. Determine the growth factor of an exponential function.
2. Identify the properties of the graph of an exponential function defined by $y = b^x$, where $b > 1$.
3. Graph an increasing exponential function.

Practice Exercises

For #1-4, complete the table for each function.

1. $f(x) = 4x$

x	-3	-2	-1	0	1	2	3
$f(x)$							

2. $f(x) = 4x^2$

x	-3	-2	-1	0	1	2	3
$f(x)$							

3. $f(x) = x^4$

x	-3	-2	-1	0	1	2	3
$f(x)$							

4. $f(x) = 4^x$

x	-3	-2	-1	0	1	2	3
$f(x)$							

5. Use your graphing calculator to sketch $f(x) = 4^x$.

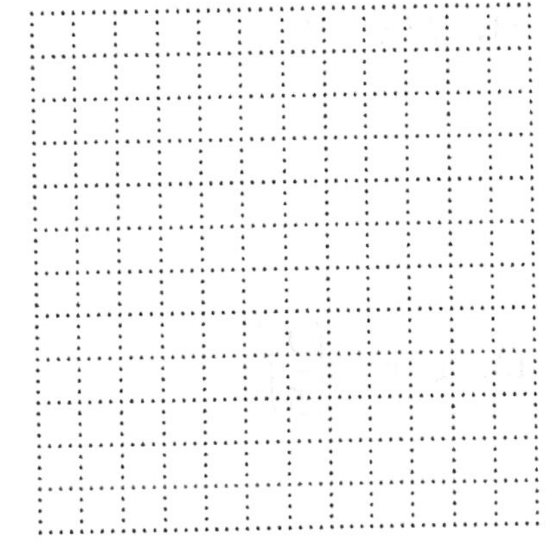

For #6-9, complete the table for each function.

6. $f(x) = 6x$

x	-3	-2	-1	0	1	2	3
$f(x)$							

7. $f(x) = 6x^2$

x	-3	-2	-1	0	1	2	3
$f(x)$							

8. $f(x) = x^6$

x	-3	-2	-1	0	1	2	3
$f(x)$							

9. $f(x) = 6^x$

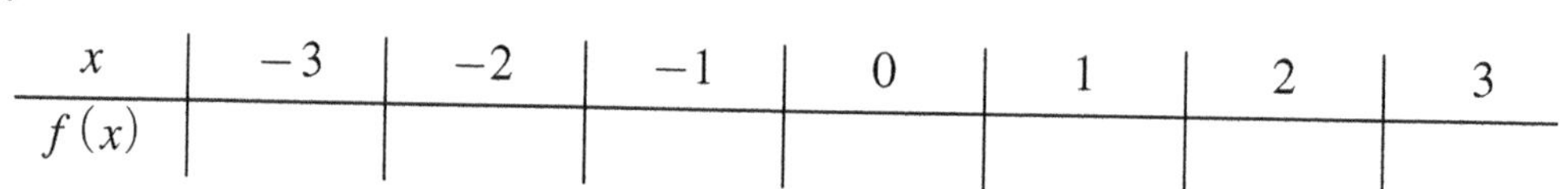

x	-3	-2	-1	0	1	2	3
$f(x)$							

10. Use your graphing calculator to sketch $f(x) = 6^x$.

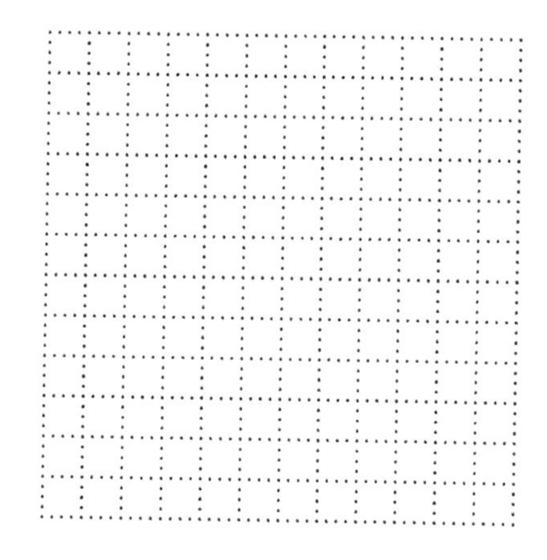

For #11-18, identify the growth factor, if any, for the function defined by the given equation.

11. $f(x) = 6^x$

12. $y = 4.2^x$

11. ______________

12. ______________

13. $g(x) = \left(\frac{9}{5}\right)^x$

14. $h(x) = \left(\frac{14}{3}\right)^x$

13. ______________

14. ______________

Name: Date:
Instructor: Section:

15. $y = 0.5^x$

16. $y = 9.9^x$

15. ______________

16. ______________

17. $w(x) = 7x$

18. $f(x) = 1.1^x$

17. ______________

18. ______________

For #19-28, consider the function $f(x) = 8^x$.

19. Complete the following table.

x	-3	-2	-1	0	1	2	3
$f(x)$							

20. Determine the domain of the function f.

21. Determine the range of the function f.

20. ______________

21. ______________

22. Is the function f increasing or decreasing?

23. Determine the growth factor.

22. ______________

23. ______________

24. What is the vertical intercept?

25. What is the horizontal intercept?

24. ______________

25. ______________

26. Explain your answer to Exercise #25.

____

____

27. Determine the horizontal asymptote.

28. Is the function continuous or discrete?

27. ______________

28. ______________

____

Concept Connections

29. Give a definition of horizontal asymptote for an exponential function $y = b^x$, where $b > 1$.

____

____

30. Why isn't 1 included for the possible values of b in $y = b^x$?

____

____

____

Name: Date:
Instructor: Section:

Chapter 3 EXPONENTIAL AND LOGORITHMIC FUNCTIONS

Activity 3.2

Learning Objectives

1. Determine the decay factor of an exponential function.
2. Graph a decreasing exponential function.
3. Identify the properties of an exponential function defined by $y = b^x$, where $b > 0$ and $b \neq 1$.

Practice Exercises

For #1-7, consider the function $f(x) = \left(\frac{3}{5}\right)^x$.

1. Use your graphing calculator to sketch the graph.

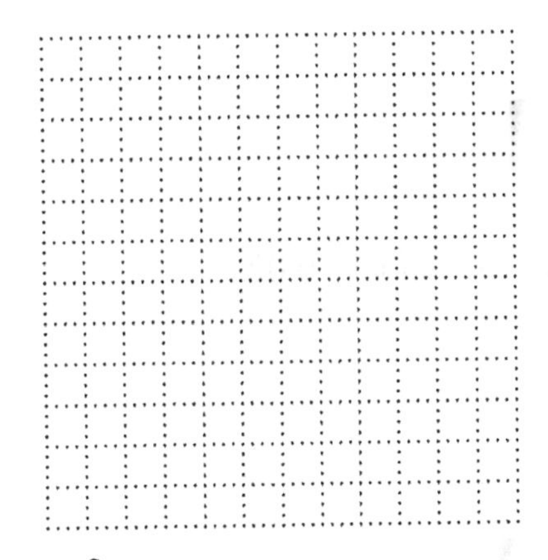

2. Is the function f increasing or decreasing?

3. What is the growth or decay factor?

4. What is the vertical intercept?

5. Determine the domain of the function f.

6. Determine the range of the function f.

7. Determine the asymptote.

2. ______________

3. ______________

4. ______________

5. ______________

6. ______________

7. ______________

For #8-15, identify the decay factor, if any, for the function defined by the given equation.

8. $f(x) = 0.6^x$ **9.** $y = 0.42^x$

8. ____________

9. ____________

10. $g(x) = \left(\frac{3}{5}\right)^x$ **11.** $h(x) = \left(\frac{4}{3}\right)^x$

10. ____________

11. ____________

12. $y = 0.05^x$ **13.** $y = 9.9^x$

12. ____________

13. ____________

14. $w(x) = 0.02x$ **15.** $f(x) = 3.01^x$

14. ____________

15. ____________

For #16-20, consider y as an exponential function of x.

16. If the decay factor is 0.2, complete the table.

x	0	1	2	3
y	0.1			

17. If the growth factor is 2.2, complete the table.

x	0	1	2	3
y	100			

18. If the growth factor is 3.5, complete the table.

x	0	1	2	3
y	8.4			

19. If the decay factor is 0.91, complete the table.

x	0	1	2	3
y	0.5			

Name: Date:
Instructor: Section:

20. If the growth factor is 1.1, complete the table.

x	0	1	2	3
y	5			

For #21-22, use the following data set.

x	−2	−1	0	1	2	3
y	−3.5	−0.5	2.5	5.5	8.5	11.5

21. Determine if the set is linear or exponential.

22. Determine the slope, growth factor, or decay factor.

21. ______________

22. ______________

For #23-24, use the following data set.

x	−2	−1	0	1	2	3
y	4.9383	2.222	1	0.45	0.2025	0.0911

23. Determine if the set is linear or exponential.

24. Determine the slope, growth factor, or decay factor.

23. ______________

24. ______________

For #25-26, use the following data set.

x	−2	−1	0	1	2	3
y	13.5	8	2.5	−3	−8.5	−14

25. Determine if the set is linear or exponential.

26. Determine the slope, growth factor, or decay factor.

25. ______________

26. ______________

For #27-28, use the following data set.

x	-2	-1	0	1	2	3
y	0.4081	1.4286	5	17.5	61.25	214.375

27. Determine if the set is linear or exponential.

28. Determine the slope, growth factor, or decay factor.

27. ______________

28. ______________

Concept Connections

29. Give a definition for the decay factor for an exponential function $y = b^x$.

____

____

30. Identify the similarities and differences between an exponential growth function and an exponential decay function with respect to domain, range, general shape, horizontal asymptote, and continuity.

____

____

Name: Date:
Instructor: Section:

Chapter 3 EXPONENTIAL AND LOGORITHMIC FUNCTIONS

Activity 3.3

Learning Objectives

1. Determine the growth and decay factor for an exponential function represented by a table of values or an equation.
2. Graph exponential functions defined by $y = ab^x$, where $b > 0$ and $b \neq 1$, $a \neq 0$.
3. Determine the doubling and halving time.

Practice Exercises

For #1-6, consider the function $f(x) = 24(3)^x$.

1. What is the y-intercept?

2. What is the growth or decay factor?

3. Use the function to determine $f(0.5)$.

4. Use your graphing calculator to estimate x when $f(x) = 48$.

5. Determine the value of x, where the value of the y-intercept will double.

6. What is the doubling time of this function?

1. ______________

2. ______________

3. ______________

4. ______________

5. ______________

6. ______________

For #7-12, consider the function $g(x) = 8(0.1)^x$.

7. What is the y-intercept?

8. What is the growth or decay factor?

7. ______________

8. ______________

9. Use the function to determine $g(0.25)$.

10. Use your graphing calculator to estimate x when $g(x) = 4$.

9. ______________

10. ______________

11. Determine the value of x, where the value of the y-intercept will be halved.

12. What is the half-life of this function?

11. ______________

12. ______________

For #13-14, consider the following table.

x	0	1	2	3	4
y	1	5	25	125	625

13. Does the relationship represent an exponential function?

14. Explain.

13. ______________

14. ______________

____

For #15-16, consider the following table.

x	1	2	3	4	5
y	1890	983	511	266	138

15. Does the relationship represent an exponential function?

16. Explain.

15. ______________

16. ______________

____

Name: Date:
Instructor: Section:

For #17-18, consider the following table.

x	0	1	2	3	4
y	0	5	10	15	20

17. Does the relationship represent an exponential function?

18. Explain.

17. ______________

18. ______________

____

For #19-20, consider the following table.

x	1	2	3	4	5
y	2	10	50	250	1250

19. Does the relationship represent an exponential function?

20. Explain.

19. ______________

20. ______________

____

For #21-24, consider the function $f(x)=5\left(\frac{1}{2}\right)^x$. *Approximate each of the following to the nearest hundredth.*

21. $f(-2)$

22. $f(2)$

21. ______________

22. ______________

23. $f\left(\frac{1}{2}\right)$

24. $f(5)$

23. ______________

24. ______________

25. Determine whether the exponential function $y = (0.95)^x$ is increasing or decreasing.

26. Explain your answer to Exercise #25.

25. ______________

26. ______________

____

27. Determine whether the exponential function $y = 4.5^x$ is increasing or decreasing.

28. Explain your answer to Exercise #27.

27. ______________

28. ______________

____

Concept Connections

29. Explain what doubling time of an exponential function means.

____

____

30. Explain what half-life of an exponential function means.

____

____

Name: Date:
Instructor: Section:

Chapter 3 EXPONENTIAL AND LOGORITHMIC FUNCTIONS

Activity 3.4

Learning Objectives

1. Determine the annual growth or decay rate of an exponential function represented by a table of values or an equation.
2. Graph an exponential function having equation $y = a(1+r)^x$.

Practice Exercises

For #1-5, determine the growth factor or growth rate.

1. Growth factor is 1.89. Find the growth rate.

2. Growth factor is 1.076. Find the growth rate.

3. Growth rate is 102%. Find the growth factor.

4. Growth factor is 5.67. Find the growth rate.

5. Growth factor is 1.298. Find the growth rate.

1. __________
2. __________
3. __________
4. __________
5. __________

For #6-11, determine the decay factor or decay rate.

6. Decay factor is 0.98. Find the decay rate.

7. Decay factor is 0.44. Find the decay rate.

8. Decay rate is 37.4%. Find the decay factor.

9. Decay rate is 0.99%. Find the decay factor.

6. __________
7. __________
8. __________
9. __________

10. Decay factor is 0.0017. Find the decay rate.

11. Decay factor is 0.6676. Find the decay rate.

10. ______________

11. ______________

For #12-17, determine the growth rate for each function.

12. $f(x) = 750(1.62)^x$

13. $g(x) = 9000(1.37)^x$

12. ______________

13. ______________

14. $h(x) = 376(1.008)^x$

15. $k(x) = 945(1.043)^x$

14. ______________

15. ______________

16. $F(x) = 190(1.090)^x$

17. $G(x) = 334(1.258)^x$

16. ______________

17. ______________

For #18-23, determine the decay rate for each function.

18. $f(x) = 324(0.65)^x$

19. $g(x) = 111(0.438)^x$

18. ______________

19. ______________

Name: Date:
Instructor: Section:

20. $h(x)=757(0.24)^x$

21. $k(x)=95(0.72)^x$

20. ____________

21. ____________

22. $F(x)=21(0.9)^x$

22. ____________

For #23-28, you have recently purchased a camper for $50,000.

23. Assuming that the value depreciates at a constant rate of 10%, write an equation that represent the value, $V(t)$, of the camper t years from now.

24. What is the decay rate of this situation?

23. ____________

24. ____________

25. What is the decay factor of this situation?

26. Use the equation to estimate the value of the camper after 10 years.

25. ____________

26. ____________

27. Use your graphing calculator to sketch the graph.

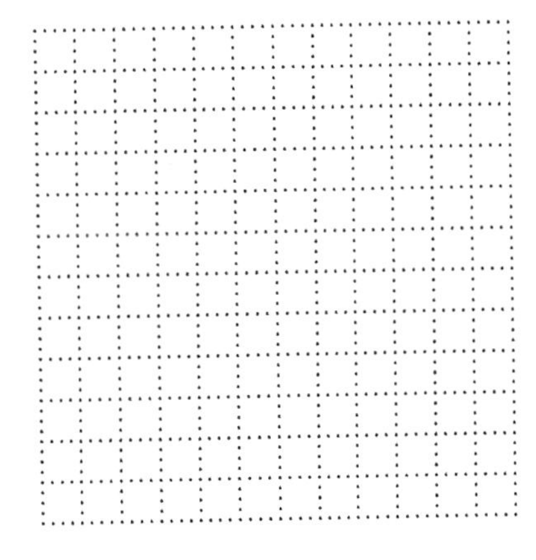

28. Use the trace and table features of your graphing calculator to determine when the camper will be worth $25,000.

28. ____________

Concept Connections

29. Explain the difference between a growth factor and a growth rate.

____

____

30. Explain the difference between a linear function and an exponential function.

____

____

Name: Date:
Instructor: Section:

Chapter 3 EXPONENTIAL AND LOGORITHMIC FUNCTIONS

Activity 3.5

Learning Objectives

1. Apply the compound interest and continuous compounding formulas to a given situation.

Key Terms

Use the vocabulary terms listed below to complete each statement in Exercises 1–4.

compound **continuous** **periods** **simple**

1. The formula for ________________ compounding is $A = Pe^{rt}$.

2. The formula for ________________ interest is $A = Prt$.

3. The formula for ________________ interest is $A = P\left(1+\frac{r}{n}\right)^{nt}$.

4. If the number of ________________ is large, $A = P\left(1+\frac{r}{n}\right)^{nt}$ is approximated by $A = Pe^{rt}$.

Practice Exercises

For #5-8, use the following scenario.

You inherit $58,000 and deposit it into an account that earns 3.5% annual interest compounded quarterly.

5. Write an equation that gives the amount of money in the account after t years.

6. How much money will be in the account after 5 years?

5. ________________

6. ________________

7. How much money will be in the account after 10 years?

8. You want to have about $100,000 in the bank when your first child begins college. Use your graphing calculator to determine in how many years you will reach this goal.

7. ________________

8. ________________

For #9-16, calculate each balance.

9. \$5000 invested at 3% compounded monthly after 7 years

10. \$12,000 invested at 4.2% compounded monthly after 3 years

9. ______________

10. ______________

11. \$21,500 invested at 1.9% compounded quarterly after 6 years

12. \$2500 invested at 5.3% compounded quarterly after 9 years

11. ______________

12. ______________

13. \$10,750 invested at 1.8% compounded continuously after 12 years

14. \$8200 invested at 3.9% compounded continuously after 8 years

13. ______________

14. ______________

15. \$4000 invested at 1.2% compounded continuously after 3 years

16. \$50,000 invested at 2.1% compounded continuously after 3 years

15. ______________

16. ______________

For #17-24, determine each effective yield. Round to the nearest thousandth of a percent.

17. APR 1.5% compounded monthly

18. APR 3.0% compounded monthly

17. ______________

18. ______________

Name: Date:
Instructor: Section:

19. APR 6.9% compounded monthly

20. APR 9.9% compounded monthly

21. APR 12.5% compounded monthly

22. APR 15.8% compounded monthly

23. APR 19.5% compounded monthly

24. APR 24.9% compounded monthly

19. ______________

20. ______________

21. ______________

22. ______________

23. ______________

24. ______________

For #25-28, use the following scenario.
You deposit $5000 into an account that earns 1.3% annual interest compounded monthly.

25. Write an equation that gives the amount of money in the account after *t* years.

26. How much money will be in the account after 3 years?

27. Estimate how long it would take for your investment to double.

28. Identify the annual growth rate and the growth factor.

25. ______________

26. ______________

27. ______________

28. ______________

Concept Connections

29. Name some common financial instruments where you could invest and collect interest.

____

____

30. Name some common financial instruments where you would have to pay interest.

____

____

Name: Date:
Instructor: Section:

Chapter 3 EXPONENTIAL AND LOGORITHMIC FUNCTIONS

Activity 3.6

Learning Objectives

1. Discover the relationship between the equations of exponential functions defined by $y = ab^t$ and the equations of continuous growth and decay exponential functions defined by $y = ae^{kt}$.
2. Solve problems involving continuous growth and decay models.
3. Graph base *e* exponential functions.

Practice Exercises

For #1-4, consider the equation $y = 17(1.48)^t$.

1. Rewrite the equation into a continuous growth equation of the form $y = ae^{kt}$.

2. What is the continuous growth rate?

1. ______________

2. ______________

3. What is the initial amount present?

4. What is the doubling time?

3. ______________

4. ______________

For #5-8, consider the equation $y = 14.6(0.44)^t$.

5. Rewrite the equation into a continuous decay equation of the form $y = ae^{kt}$.

6. What is the continuous decay rate?

5. ______________

6. ______________

7. What is the initial amount present?

8. What is the doubling time?

7. ______________

8. ______________

For #9-12, consider the exponential function $P = 7150e^{0.03t}$.

9. Is the function increasing or decreasing?

10. Determine the continuous growth or decay rate.

9. ________________

10. ________________

11. What is the initial value?

12. Determine the doubling time.

11. ________________

12. ________________

For #13-16, consider the exponential function $Q = 1300(0.79)^t$.

13. Is the function increasing or decreasing?

14. Determine the continuous growth or decay rate.

13. ________________

14. ________________

15. What is the initial value?

16. Determine the half-life.

15. ________________

16. ________________

For #17-20, consider the exponential function $A = 92(1.091)^t$.

17. Is the function increasing or decreasing?

18. Determine the continuous growth or decay rate.

17. ________________

18. ________________

19. What is the initial value?

20. Determine the doubling time.

19. ________________

20. ________________

Name: Date:
Instructor: Section:

For #21-24, consider the exponential function $R = 31e^{-0.16t}$.

21. Is the function increasing or decreasing?

22. Determine the continuous growth or decay rate.

23. What is the initial value?

24. Determine the half-life.

21. ______________

22. ______________

23. ______________

24. ______________

For #25-28, consider the exponential function $S = 0.737(0.94)^t$.

25. Is the function increasing or decreasing?

26. Determine the continuous growth or decay rate.

27. What is the initial value?

28. Determine the half-life.

25. ______________

26. ______________

27. ______________

28. ______________

Concept Connections

29. What kind of number does e represent? What is the approximate value of e?

____

____

30. For an exponential function with base e, will that function be increasing, decreasing, or unknown without further information?

____

____

Name: Date:
Instructor: Section:

Chapter 3 EXPONENTIAL AND LOGORITHMIC FUNCTIONS

Activity 3.7

Learning Objectives

1. Determine the regression equation of an exponential function that best fits the given data.
2. Make predictions using an exponential regression equation.
3. Determine whether a linear or exponential model best fits the data.

Practice Exercises

For #1-11, use the data in the following table.

x	4	7	10	14	19
y	60.5	29.7	14.6	5.7	1.7

1. Use your graphing calculator to sketch the scatterplot.

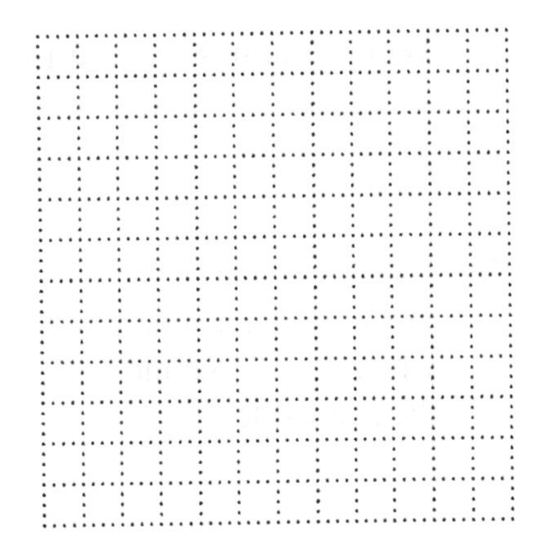

2. Use your graphing calculator to determine the regression equation of a linear function.

2. 

3. Use your graphing calculator to sketch the scatterplot with the linear regression function.

4. Use your graphing calculator to determine the regression equation of an exponential function.

4. ______________

5. Use your graphing calculator to sketch the scatterplot with the linear and exponential regression functions.

6. Which graph is the best fit?

6. ______________

7. Use the exponential model to determine y when $x = 12$. Round to the nearest tenth.

8. Use the exponential model to determine y when $x = 25$. Round to the nearest tenth.

7. ______________

8. ______________

9. According to the exponential mode, what is the decay factor?

10. What is the half-life for the exponential model?

9. ______________

10. ______________

11. Does the graph have a horizontal asymptote?

11. ______________

Name: Date:
Instructor: Section:

For #12-22, use the data in the following table.

x	1	5	8	12	15
y	87	162	271	548	916

12. Use your graphing calculator to sketch the scatterplot.

13. Use your graphing calculator to determine the regression equation of a linear function.

13. 

14. Use your graphing calculator to sketch the scatterplot with the linear regression function.

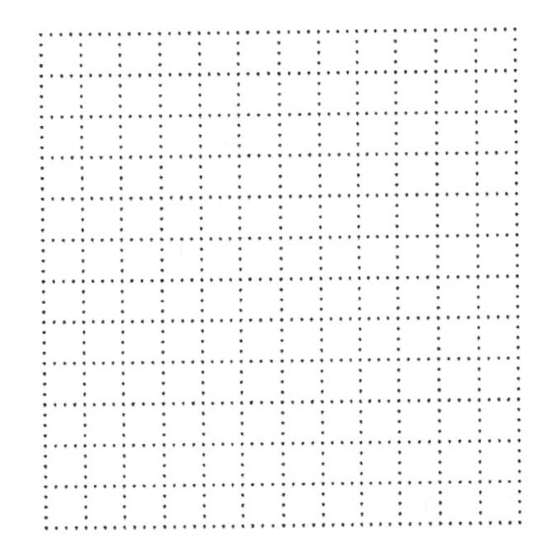

15. Use your graphing calculator to determine the regression equation of an exponential function.

15. ________________

16. Use your graphing calculator to sketch the scatterplot with the linear and exponential regression functions.

17. Which graph is the best fit?

17. ______________

18. Use the exponential model to determine y when $x = 10$.

19. Use the exponential model to determine y when $x = 20$.

18. ______________

19. ______________

20. According to the exponential mode, what is the growth factor?

21. What is the half-life for the exponential model?

20. ______________

21. ______________

22. Does the graph have a horizontal asymptote?

22. ______________

For #23-28, consider the exponential function $y = 4(3)^x$.

23. What is the domain?

24. What is the range?

23. ______________

24. ______________

Name: Date:
Instructor: Section:

25. For what value of x is the exponential function positive?

26. For what value of x is the exponential function negative?

25. ______________

26. ______________

27. What is the vertical intercept?

28. What is the doubling time?

27. ______________

28. ______________

Concept Connections

29. Use the function $y = 2510(1.16)^x$ and your graphing calculator to determine the value of x when y = 30,000. Round to the nearest tenth.

____

____

30. Use the function $y = 45,220(0.79)^x$ and your graphing calculator to determine the value of x when y = 12,000. Round to the nearest tenth.

____

____

Name: Date:
Instructor: Section:

Chapter 3 EXPONENTIAL AND LOGORITHMIC FUNCTIONS

Activity 3.8

Learning Objectives

1. Define *logarithm.*
2. Write an exponential statement in logarithmic form.
3. Write a logarithmic statement in exponential form.
4. Determine log and ln values using a calculator.

Key Terms

Use the vocabulary terms listed below to complete each statement in Exercises 1–4.

common **exponential** **logarithmic** **natural**

1. The equation $\log_5 x = 25$ is an example of a(n) ________________ equation.

2. Base-10 logarithms are called ________________ logarithms.

3. Base e logarithms are called ________________ logarithms.

4. The equation $3^x = 10$ is an example of a(n) ________________ equation.

Practice Exercises

For #5-12, use the definition of logarithms to determine the exact value of each of the following expressions.

5. $\log 10^6$

6. $\log_8 \left(\dfrac{1}{64}\right)$

5. ________________

6. ________________

7. $\log e^9$

8. $\ln 1$

7. ________________

8. ________________

9. $\ln\left(\frac{1}{e^5}\right)$

10. $\log_5(125)$

9. ______________

10. ______________

11. $\log_2 \sqrt[3]{2}$

12. $\log 1$

11. ______________

12. ______________

For #13-16, rewrite the following equation in logarithmic form.

13. $27^{2/3} = 9$

14. $d^t = m$

13. ______________

14. ______________

15. $e^{-3} = 0.0498$

16. $\sqrt{144} = 12$

15. ______________

16. ______________

For #17-20, rewrite the following equation in exponential form.

17. $\log_5 0.2 = -1$

18. $\log_a t = p$

17. ______________

18. ______________

19. $\ln 0.4 = -0.9163$

20. $\log\sqrt{10} = \frac{1}{2}$

19. ______________

20. ______________

Name: Date:
Instructor: Section:

For #21-28, use your calculator to evaluate the following. Round your answer to four decimal places.

21. $\log 7$

22. $\ln 0.1$

21. ____________

22. ____________

23. $\log 8$

24. $\ln 8$

23. ____________

24. ____________

25. $\log 0.3$

26. $\ln 15$

25. ____________

26. ____________

27. $\log \frac{3}{4}$

28. $\ln \frac{3}{4}$

27. ____________

28. ____________

Concept Connections

29. For the logarithmic function $f(x) = \log_b x$, what is the domain and range?

____

____

30. For the logarithmic function $f(x) = \log_b x$, what does b represent, and what are the limitations of b?

____

____

Name: Date:
Instructor: Section:

Chapter 3 EXPONENTIAL AND LOGORITHMIC FUNCTIONS

Activity 3.9

Learning Objectives

1. Determine the inverse of the exponential function.
2. Identify the properties of the graph of a logarithmic function.
3. Graph the natural logarithmic function.

Practice Exercises

For #1-5, determine the equation of the inverse of each function.

1. $y = 2^x$

2. $y = 12^x$

3. $y = \left(\frac{1}{2}\right)^x$

4. $y = (0.1)^x$

5. $y = 100^x$

1. ________________

2. ________________

3. ________________

4. ________________

5. ________________

For #6-17, consider the function $y = 5\ln x$.

6. Complete the table.

x	0	1	5	10	20
y					

7. Use your graphing calculator to sketch the graph.

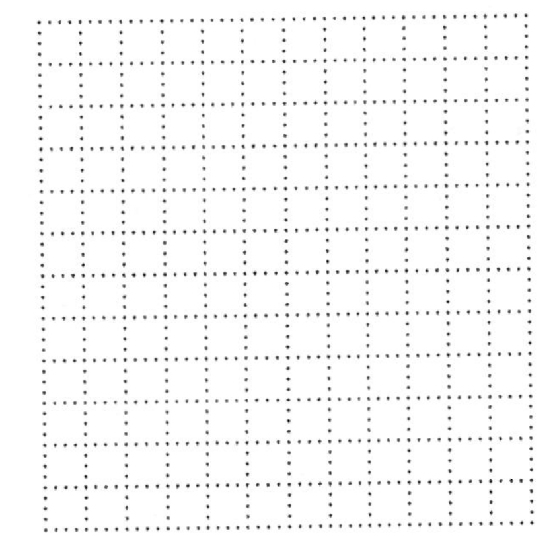

8. What is the domain of the function?

9. What is the range of the function?

8. ______________

9. ______________

10. For what values of x is the graph increasing?

11. What is the x-intercept?

10. ______________

11. ______________

12. What is the y-intercept?

13. Does the graph have a horizontal asymptote?

12. ______________

13. ______________

____

14. Complete the table.

x	1	0.1	0.01	0.001
y				

15. As the input values take on values closer and closer to 0, what happens to the corresponding output values?

16. Determine the vertical asymptote.

15. ______________

____

16. ______________

17. What is the equation of the vertical asymptote of the graph $y = 5\ln(x+5)$.

17. ______________

Name: Date:
Instructor: Section:

For #18-28, consider the function $y = \log(5x)$.

18. Complete the table.

x	0	1	5	10	20
y					

19. Use your graphing calculator to sketch the graph.

20. What is the domain of the function?

21. What is the range of the function?

20. ______________

21. ______________

22. For what values of x is the graph decreasing?

23. What is the x-intercept?

22. ______________

23. ______________

24. Does the graph have a horizontal asymptote?

24. ______________

____

25. Complete the table.

x	1	0.1	0.01	0.001
y				

26. As the input values take on values closer and closer to 0, what happens to the corresponding output values?

27. Determine the vertical asymptote.

26. ____________

____

27. ____________

28. What is the equation of the vertical asymptote of the graph $y = \log(5x - 5)$.

28. ____________

Concept Connections

29. For the function defined by $f(x) = 10^x$, what is its inverse?

____

____

30. What is the difference between $f(x) = \log x$ and $g(x) = \ln x$?

____

____

Name: Date:
Instructor: Section:

Chapter 3 EXPONENTIAL AND LOGORITHMIC FUNCTIONS

Activity 3.10

Learning Objectives

1. Compare the average rate of change of increasing logarithmic, linear, and exponential functions.
2. Determine the regression equation of a natural logarithmic function having equation $y = a + b \ln x$ that best fits a set of data.

Practice Exercises

For #1-11, use the data in the following table.

x	2	5	8	11	14
y	7.59	9.47	10.43	11.08	11.58

1. Use your graphing calculator to sketch the scatterplot.

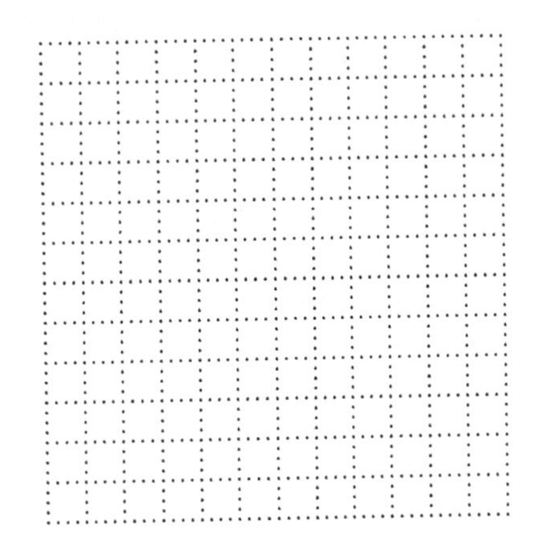

2. Use your graphing calculator to determine the regression equation of a linear function.

2. 

3. Use your graphing calculator to sketch the scatterplot with the linear regression function.

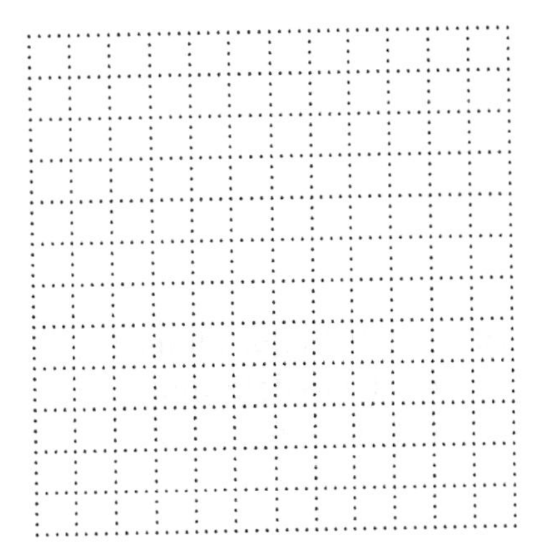

4. Use your graphing calculator to determine the regression equation of an exponential function.

4. ______________

5. Use your graphing calculator to sketch the scatterplot with the linear and exponential regression functions.

6. Use your graphing calculator to determine the regression equation of a logarithmic function.

6. ______________

7. Use your graphing calculator to sketch the scatterplot with the linear, exponential, and logarithmic regression functions.

8. Which graph is the best fit?

9. Explain your answer to Exercise #8.

8. ______________

9. ______________

____

10. Use the logarithmic model to determine y when $x = 6$.

11. Use the logarithmic model to determine y when $x = 20$.

10. ______________

11. ______________

Name: Date:
Instructor: Section:

For #12-20, use the data in the following table.

x	5	10	75	125
y	3.24	4.37	7.67	8.50

12. Determine the average rate of change of y as x increases from 5 to 10.

13. Determine the average rate of change of y as x increases from 10 to 75.

12. ______________

13. ______________

14. Determine the average rate of change of y as x increases from 75 to 125.

15. What can you say in general about the average rate of change?

14. ______________

15. ______________

____

16. Use your graphing calculator to sketch the scatterplot.

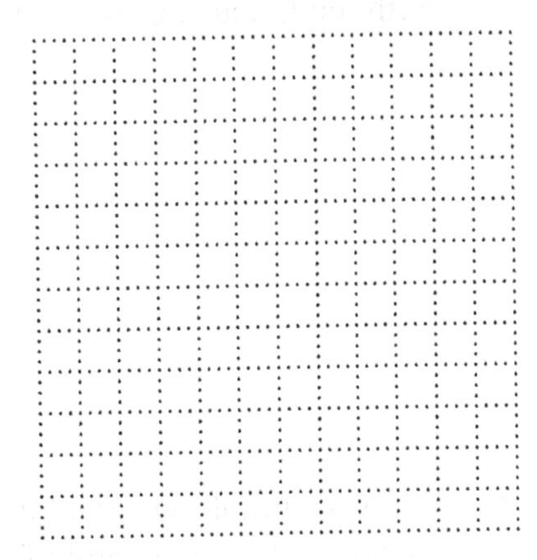

17. Do the table and scatterplot indicate that the data is linear, exponential or logarithmic? Explain.

18. Use your graphing calculator to determine the regression equation of a logarithmic function.

17. ______________

____

18. ______________

19. Use the logarithmic model to determine y when $x = 100$.

20. Use the logarithmic model to determine y when $x = 200$.

19. ______________

20. ______________

For #21-28, use the data in the following table.

x	2	4	6	8	10
y	13.15	19.59	23.86	26.05	28.12

21. Use your graphing calculator to sketch the scatterplot.

22. Use your graphing calculator to determine the regression equation of a linear function.

22. ____________________

23. Use your graphing calculator to sketch the scatterplot with the linear regression function.

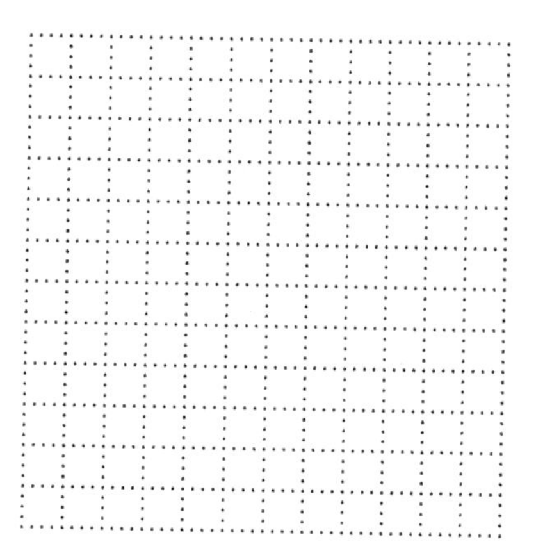

24. Use your graphing calculator to determine the regression equation of a logarithmic function.

24. ____________________

25. Use your graphing calculator to sketch the scatterplot with the linear and logarithmic regression functions.

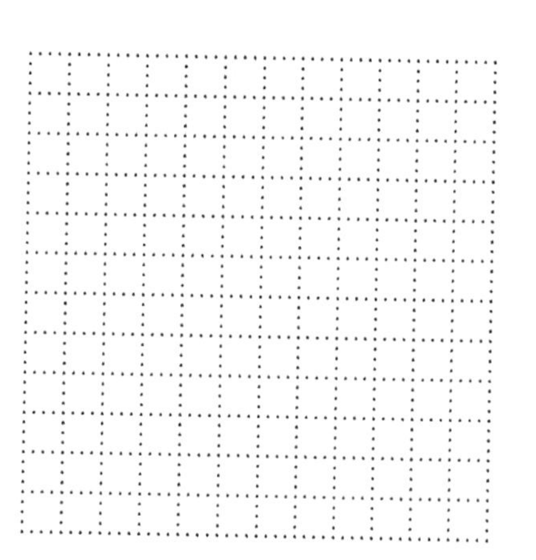

Name:
Instructor:

Date:
Section:

26. Which graph has the best fit?

26. ______________

27. Use the logarithmic model to determine y when $x = 5$. Round to the nearest tenth.

28. Use the logarithmic model to determine y when $x = 13$. Round to the nearest tenth.

27. ______________

28. ______________

Concept Connections

29. Describe the rate at which the output values change for an increasing linear function, an increasing exponential function and an increasing logarithmic function.

____

____

____

30. For an increasing linear function, an increasing exponential function and an increasing logarithmic function, describe or compare the growth rate in general terms.

____

____

Name: Date:
Instructor: Section:

Chapter 3 EXPONENTIAL AND LOGORITHMIC FUNCTIONS

Activity 3.11

Learning Objectives

1. Apply the log of a product property.
2. Apply the log of a quotient property.
3. Apply the log of a power property.
4. Discover the change of base formula.

Practice Exercises

For #1-4, use the properties of logarithms to write the following as a sum or difference of two or more logarithms.

1. $\log_b(6\cdot 216)$

2. $\log_8(2\cdot 15)$

3. $\log_{13}\frac{11}{19}$

4. $\log_5\frac{xy}{6}$

1. ______________

2. ______________

3. ______________

4. ______________

For #5-8, write the following expressions as a single logarithm.

5. $\log_a 10+\log_a 9$

6. $\log_7 A+\log_7 B$

7. $\log_a(x+1)-\log_a 11$

8. $\log_c 48-\log_c 4$

5. ______________

6. ______________

7. ______________

8. ______________

For #9-12, given that $\log_a x = 12$ and that $\log_a y = 3$, determine the numeric value of each of the following.

9. $\log_a y^2$ **10.** $\log_a \sqrt{x}$

9. ______________

10. ______________

11. $5 + \log_a x^3$ **12.** $\log_a \frac{x}{ay}$

11. ______________

12. ______________

For #13-16, use the properties of logarithms to write the given logarithms as the sum or difference of two or more logarithms or as the product of a real number and a logarithm. Simplify, if possible.

13. $\log_b x^2 yz^3$ **14.** $\log_5 5^8$

13. ______________

14. ______________

15. $\log_7 7^x$ **16.** $\ln \frac{w\sqrt[5]{x}}{y^3 z^4}$

15. ______________

16. ______________

For #17-20, write each of the following as the logarithm of a single expression with coefficient 1.

17. $2\log_3 2 + \log_3 5$ **18.** $\frac{1}{3}\log w^4 - \frac{1}{3}\log y^7$

17. ______________

18. ______________

Name:
Instructor:

Date:
Section:

19. $5\ln 3 - 2\ln 6 + 6\ln x$

20. $\log_4(x-3) - 2\log_4(x+1)$

19. ______________

20. ______________

For #21-28, use the change of base formula and your calculator to determine a decimal approximation of each of the following to the nearest ten-thousandth.

21. $\log_4 85$

22. $\log_5 100$

21. ______________

22. ______________

23. $\log_{0.2} 90$

24. $\log_3 32$

23. ______________

24. ______________

25. $\log_7 78$

26. $\log_2 \sqrt{5}$

25. ______________

26. ______________

27. $\log_{15} 125$

28. $\log_{0.8} \sqrt[3]{41}$

27. ______________

28. ______________

Concept Connections

29. Sam incorrectly thinks that $\log_b(x+y)$ is equivalent to $\log_b x + \log_b y$. How can you convince Sam that the two expressions are not equivalent?

____

____

30. Amelia incorrectly thinks that $\log_2 5 + \log_3 5$ can be written as $\log_2 25$. What is Amelia doing wrong?

____

____

Name: Date:
Instructor: Section:

Chapter 3 EXPONENTIAL AND LOGORITHMIC FUNCTIONS

Activity 3.12

Learning Objectives

1. Solve exponential equations both graphically and algebraically.

Practice Exercises

For #1-20, solve each equation using an algebraic approach. Round to four decimal places.

1. $2^x = 11$

2. $8 = 2^{x-1}$

3. $e^t = 200$

4. $3.6^x = 25$

5. $5^{3x} = 125$

6. $4e^{3x} = 10$

7. $3^x = 10$

8. $4^{2x-1} = 64$

1. ______________

2. ______________

3. ______________

4. ______________

5. ______________

6. ______________

7. ______________

8. ______________

9. $4^{x+3}=3^{x-1}$

10. $e^{-0.05t}=0.09$

9. ______________

10. ______________

11. $3^{4x+3}=243$

12. $8^{3x}=24$

11. ______________

12. ______________

13. $12,000=8000(1.04)^{t}$

14. $6000=5000(1.03)^{t}$

13. ______________

14. ______________

15. $e^{-0.8t}=3$

16. $e^{4.6t}=5$

15. ______________

16. ______________

17. $e^{0.03t}=1.5$

18. $e^{-0.05t}=0.5$

17. ______________

18. ______________

Name: Date:
Instructor: Section:

19. $14 = 9.6e^{0.02t}$

20. $22 = 4e^{1.79t}$

19. ______________

20. ______________

For #21-28, solve each equation graphically. Round to four decimal places.

21. $300 = 200(1.627)^t$

22. $160 = 110(1.108)^t$

21. ______________

22. ______________

23. $11^{5x+2} = 12$

24. $8^{2x-3} = 9$

23. ______________

24. ______________

25. $e^{0.001t} = 4$

26. $e^{-0.07t} = 4$

25. ______________

26. ______________

27. $20{,}000 = 16{,}000(1.07)^t$

28. $6000 = 4500(1.03)^t$

27. ______________

28. ______________

Concept Connections

29. How old is an archaeological discovery that has lost 15% of its carbon-14? Use $P(t) = P_0 e^{-0.00012t}$.

____

____

30. A college loan of \$58,000 is made at 3% interest, compounded annually. After t years, the amount due, A is given by the function $A(t) = 58{,}000(1.03)^t$. After how many years will the amount due reach \$73,400?

Name: Date:
Instructor: Section:

Chapter 4 QUADRATIC AND HIGHER-ORDER POLYNOMIAL FUNCTIONS

Activity 4.1

Learning Objectives

1. Identify the functions of the form $f(x) = ax^2 + bx + c$ as quadratic functions.
2. Explore the role of *c* as it relates to the graph of $f(x) = ax^2 + bx + c$.
3. Explore the role of *a* as it relates to the graph of $f(x) = ax^2 + bx + c$.
4. Explore the role of *b* as it relates to the graph of $f(x) = ax^2 + bx + c$.

Note: $a \neq 0$ in Objectives 1–4.

Key Terms

Use the vocabulary terms listed below to complete each statement in Exercises 1–2.

parabola **quadratic**

1. The equation of a __________________ function has standard form $f(x) = ax^2 + bx + c$.

2. The graph of the function $f(x) = ax^2 + bx + c$ is called a __________________ .

Practice Exercises

For #3-4, complete the table for the given quadratic function.

3. $f(x) = -2x^2$

x	-3	-2	-1	0	1	2	3
$f(x)$							

4. $g(x) = 2x^2$

x	-3	-2	-1	0	1	2	3
$g(x)$							

For #5-8, identify the value of a, b, and c.

5. $y = 4x^2$

6. $y = 0.75x^2 - 6$

5. ______________

____

6. ______________

____

7. $y = x^2 - 5x$

8. $y = -2x^2 + 3x - 1$

7. ______________

____

8. ______________

____

For #9-10, predict what the graph of each of the following quadratic functions will look like. Use your graphing calculator to verify your prediction.

9. $f(x) = -3x^2 - 7$

10. $g(x) = 0.2x^2 + 4$

9. ______________

____

10. ______________

____

For #11-14, graph the following pair of functions on your graphing calculator, and describe any similarities as well as any differences that you observe in the graphs.

11. $f(x) = 5x^2 - 7,$
$g(x) = 5x^2 + 7$

12. $h(x) = 0.3x^2,$
$k(x) = 0.3x^2 - 4$

11. ______________

____

12. ______________

____

13. $g(x) = 4x^2,$
$h(x) = 0.4x^2$

14. $f(x) = 0.1x^2,$
$g(x) = -0.1x^2$

13. ______________

____

14. ______________

____

Name: Date:
Instructor: Section:

For #15-16, use the function $f(x) = -6x^2 + 3x - 1$.

15. Determine whether the parabola opens upward or downward.

16. Determine the vertical intercept.

15. ______________

16. ______________

For #17-18, use the function $g(x) = 2x^2 + 6$.

17. Determine whether the parabola opens upward or downward.

18. Determine the vertical intercept.

17. ______________

18. ______________

For #19-20, use the function $h(x) = 4x^2 + x$.

19. Determine whether the parabola opens upward or downward.

20. Determine the vertical intercept.

19. ______________

20. ______________

For #21-22, use the function $k(x) = -3x^2$.

21. Determine whether the parabola opens upward or downward.

22. Determine the vertical intercept.

21. ______________

22. ______________

For #23-24, determine which graph is wider.

23. $g(x) = 8x^2$, $h(x) = 0.8x^2$

24. $f(x) = x^2$, $k(x) = 3x^2$

23. ______________

24. ______________

For #25-28, for each function, determine whether the turning point is on the y-axis.

25. $f(x) = -6x^2 + 3x - 1$ **26.** $g(x) = 2x^2 + 6$

25. ____________

26. ____________

27. $h(x) = 4x^2 + x$ **28.** $k(x) = -3x^2$

27. ____________

28. ____________

Concept Connections

29. For a quadratic function $f(x) = ax^2 + bx + c$, explain the effect a has on the graph. Be sure to include the direction the parabola opens and the effect on the width of the parabola.

____

____

30. For a quadratic function $f(x) = ax^2 + bx + c$, explain the effect b has on the graph when $b = 0$ and when $b \neq 0$.

____

____

Name: Date:
Instructor: Section:

Chapter 4 QUADRATIC AND HIGHER-ORDER POLYNOMIAL FUNCTIONS

Activity 4.2

Learning Objectives

1. Determine the vertex or turning point of a parabola.
2. Identify the vertex as a maximum or minimum.
3. Determine the axis of symmetry of a parabola.
4. Identify the domain and range.
5. Determine the y-intercept of a parabola.
6. Determine the x-intercept(s) of a parabola.
7. Interpret the practical meaning of the vertex and intercepts in a given problem.

Practice Exercises

For #1-7, for each function, determine the direction in which the graph opens and the axis of symmetry.

1. $f(x)=x^2+5$

2. $g(x)=x^2+10x+27$

3. $h(x)=x^2-4x+3$

4. $h(x)=x^2+5x+7$

5. $g(x)=-x^2+8x-13$

6. $y=-2x^2+5x+4$

7. $f(x)=-x^2+x-9$

1. ____________

2. ____________

3. ____________

4. ____________

5. ____________

6. ____________

7. ____________

For #8-14, for each function, determine the turning point (vertex); determine if a maximum or minimum, and the y-intercept.

8. $f(x) = x^2 + 5$

9. $g(x) = x^2 + 10x + 27$

8. ________________

9. ________________

10. $h(x) = x^2 - 4x + 3$

11. $h(x) = x^2 + 5x + 7$

10. ________________

11. ________________

12. $g(x) = -x^2 + 8x - 13$

13. $y = -2x^2 + 5x + 4$

12. ________________

13. ________________

14. $f(x) = -x^2 + x - 9$

14. ________________

For #15-21, for each function, determine the coordinates of the x-intercepts, if they exist, and the domain and range.

15. $g(x) = -x^2 + 5x - 6$

16. $h(x) = 5x^2 + 10x + 8$

15. ________________

16. ________________

17. $y = 2x^2 - 6x + 4$

18. $f(x) = -5x^2 + 10x$

17. ________________

18. ________________

Name: Date:
Instructor: Section:

19. $y = 3x^2 + 9x + 6$

20. $h(x) = -x^2 + 2x - 1$

19. ______________

20. ______________

21. $h(x) = -x^2 - 7x - 6$

21. ______________

For #22-28, for each function, determine the horizontal interval over which each function is increasing and the horizontal interval over which each function is decreasing.

22. $g(x) = -x^2 + 5x - 6$

23. $h(x) = 5x^2 + 10x + 8$

22. ______________

23. ______________

24. $y = 2x^2 - 6x + 4$

25. $f(x) = -5x^2 + 10x$

24. ______________

25. ______________

26. $y = 3x^2 + 9x + 6$

27. $h(x) = -x^2 + 2x - 1$

26. ______________

27. ______________

28. $h(x) = -x^2 - 7x - 6$

28. ______________

Concept Connections

29. Define the axis of symmetry for a quadratic function.

____

____

30. Define the vertex for a quadratic function.

____

____

Name: Date:
Instructor: Section:

Chapter 4 QUADRATIC AND HIGHER-ORDER POLYNOMIAL FUNCTIONS

Activity 4.3

Learning Objectives
1. Solve quadratic equations numerically.
2. Solve quadratic equations graphically.
3. Solve quadratic inequalities graphically.

Practice Exercises

For #1-8, solve the quadratic equation numerically (using tables of x- and y-values). Verify your answers graphically.

1. $21 = x^2 - 4x$

2. $x^2 + 11x + 24 = 0$

3. $3x^2 = 21x + 132$

4. $x^2 - 3x - 40 = 0$

5. $2x^2 + 4x = 70$

6. $5x^2 = 30x$

7. $4x^2 = 100$

8. $x^2 - 3x - 8 = 0$

1. ______________

2. ______________

3. ______________

4. ______________

5. ______________

6. ______________

7. ______________

8. ______________

For #9-16, solve the quadratic equation graphically using at least two different approaches. When necessary, give your answers to the nearest hundredth.

9. $x^2 + 15x + 26 = 0$

10. $7x^2 - 14x - 19 = 0$

9. ______________

10. ______________

11. $25x^2 = 400$

12. $x^2 + x = 10$

11. ______________

12. ______________

13. $x^2 + 4x = 45$

14. $8x^2 = 10x$

13. ______________

14. ______________

15. $2x^2 = 11x + 40$

16. $2x^2 + 63 = 23x$

15. ______________

16. ______________

For #17-22, solve the quadratic equation by using either a numerical or a graphical approach.

17. $x^2 - x - 30 = 0$

18. $x^2 + 13x + 36 = 0$

17. ______________

18. ______________

Name: Date:
Instructor: Section:

19. $x^2 - 4x - 50 = 2x + 22$

20. $x^2 - 6x + 10 = 4x - 14$

21. $x^2 + 10x - 15 = 5x + 9$

22. $x^2 - 12x + 6 = 5x + 6$

19. ______________

20. ______________

21. ______________

22. ______________

For #23-28, solve each inequality using a graphing approach.

23. $x^2 - 4x - 5 < 16$

24. $x^2 - 4x - 5 > 16$

25. $4x^2 + 17x + 10 \leq 25$

26. $4x^2 + 17x + 10 > 25$

27. $-x^2 + 6x + 16 \leq 0$

28. $-x^2 + 6x + 16 > 0$

23. ______________

24. ______________

25. ______________

26. ______________

27. ______________

28. ______________

Concept Connections

29. Explain how to solve $f(x) = c$ graphically where $f(x)$ is a quadratic function.

____

____

30. Explain how to solve $f(x) > c$ graphically where $f(x)$ is a quadratic function.

____

____

Name: Date:
Instructor: Section:

Chapter 4 QUADRATIC AND HIGHER-ORDER POLYNOMIAL FUNCTIONS

Activity 4.4

Learning Objectives

1. Factor expressions by removing the greatest common factor.
2. Factor trinomials using trial and error.
3. Use the Zero-Product Property to solve equations.
4. Solve quadratic equations by factoring.

Practice Exercises

For #1-4, factor the polynomials by removing the GCF (greatest common factor).

1. $35x^4 - 28x^6$ **2.** $40x^8y^2 - 55x^3y^3$

1. ______________

2. ______________

3. $6x^4 - 9x^3 + 36x^2$ **4.** $24x^5 - 28x^4 + 4x^3$

3. ______________

4. ______________

For #5-16, completely factor the polynomials. Remember to look first for the GCF.

5. $x^2 + 4x - 21$ **6.** $p^2 - 18p + 65$

5. ______________

6. ______________

7. $x^2 + 3xy - 28y^2$ **8.** $x^2 - x - 72$

7. ______________

8. ______________

9. $27+12x+x^2$ **10.** $3x^2+7x-20$

9. ______________

10. ______________

11. $4x^2+17x-15$ **12.** $6x^4-19x^3-20x^2$

11. ______________

12. ______________

13. $18b^4+69b^3-105b^2$ **14.** $15x^6-95x^5+100x^4$

13. ______________

14. ______________

15. $16x^4+4x^3-30x^2$ **16.** $80x^2-296x+56$

15. ______________

16. ______________

For #17-28, solve each quadratic equation by factoring.

17. $x^2-9x+20=0$ **18.** $x^2+4x-12=0$

17. ______________

18. ______________

19. $x^2-2x=35$ **20.** $x^2-5x=24$

19. ______________

20. ______________

Name: Date:
Instructor: Section:

21. $4x^2 + 11x - 3 = 0$

22. $6x^2 + 30x = 0$

23. $x^2 - 3x = 108$

24. $5x(x-9) - 4(x-9) = 0$

25. $7x(2x+5) + 3(2x+5) = 0$

26. $9x(7x-4) + (7x-4) = 0$

27. $3x^2 = 8x$

28. $3x^2 + 14x = 5$

21. ______________

22. ______________

23. ______________

24. ______________

25. ______________

26. ______________

27. ______________

28. ______________

Concept Connections

29. State the Zero-Product Property.

____

____

30. Explain how to solve equations by factoring.

Name: Date:
Instructor: Section:

Chapter 4 QUADRATIC AND HIGHER-ORDER POLYNOMIAL FUNCTIONS

Activity 4.5

Learning Objectives

1. Solve quadratic equations by the quadratic formula.

Practice Exercises

For #1-4, identify the values of a, b, and c, and substitute into the quadratic formula. Do not solve.

1. $x^2+7x+5=0$

2. $2x^2-6x+7=0$

1. ______________

____

2. ______________

____

3. $(3x-1)(2x+6)=1$

4. $(x-5)^2+x^2=4$

3. ______________

____

4. ______________

____

For #5-26, solve each quadratic equation using the quadratic formula. Round to the nearest hundredth.

5. $x^2+7x+2=0$

6. $x^2+8x+8=0$

5. ______________

6. ______________

7. $x^2-5x+3=0$

8. $x^2-5x-8=0$

7. ______________

8. ______________

9. $x^2+3x-2=0$

10. $x^2-5x+5=0$

11. $4x^2+12x-5=0$

12. $2x^2-x-3=0$

13. $4x^2+4x-63=0$

14. $x^2-9x+16=0$

15. $x^2+4x-9=0$

16. $2x^2-3x-10=0$

17. $x^2+4x+3=0$

18. $x^2+4x-3=0$

19. $x^2-x=4$

20. $x^2+9x=-3x+16$

9. ______________

10. ______________

11. ______________

12. ______________

13. ______________

14. ______________

15. ______________

16. ______________

17. ______________

18. ______________

19. ______________

20. ______________

Name: Date:
Instructor: Section:

21. $(x+3)(x-2)=3$

22. $(3x-4)(x+3)=1$

21. ______________

22. ______________

23. $(x-1)^2+x^2=30$

24. $(x+3)^2+x^2=24$

23. ______________

24. ______________

25. $32x^2-4x-15=0$

26. $25x^2-230x+448=0$

25. ______________

26. ______________

For #27-28, find the distance from the axis of symmetry to either x-intercept.

27. $x^2-2x-8=0$

28. $x^2+14x+33=0$

27. ______________

28. ______________

Concept Connections

29. For a quadratic equation of the form $ax^2+bx+c=0$, state the quadratic formula.

____

____

30. For a quadratic equation of the form $ax^2 + bx + c = 0$, what is the distance from the axis of symmetry to either x-intercept?

Name: Date:
Instructor: Section:

Chapter 4 QUADRATIC AND HIGHER-ORDER POLYNOMIAL FUNCTIONS

Activity 4.6

Learning Objectives
1. Determine quadratic regression models using a graphing calculator.
2. Solve problems using quadratic regression models.

Practice Exercises

For #1-6, use the following data set. Round answers to three decimal places.

x	0	2	4	6	8
y	4	14	38	79	157

1. Determine an appropriate scale, and plot these points.

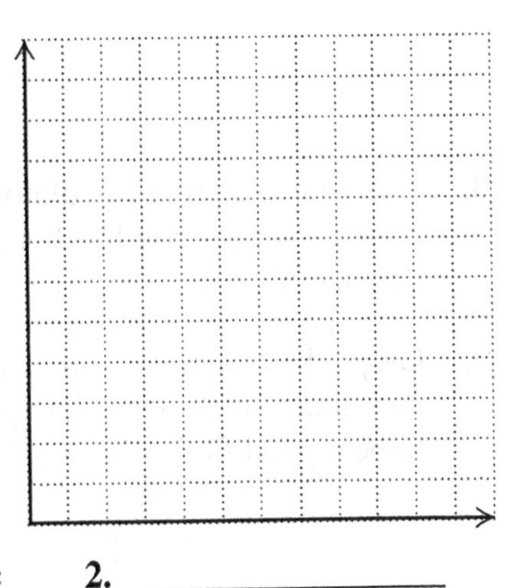

2. Use your graphing calculator to determine the quadratic regression equation for this data set.

2. ____________

3. Use your graphing calculator to graph the regression equation and data points. Use the same scale as used in Exercise #1.

4. Predict the output for $x = 5$.

5. Predict the output for $x = 7$.

4. ____________

5. ____________

6. Predict the output for $x = 10$.

6. ____________

For #7-14, use the following scenario.

An arrow is launched with a heavy bow. The height, h, of the arrow above the ground in yards as a function of time, t, in seconds, can be partially modeled by the following table.

t	0	1	2	3	4	5
$h(t)$	2	84	133	151	137	91

7. Sketch a scatterplot of the data using your graphing calculator.

8. Use your graphing calculator to determine the quadratic regression equation for this data set. Round to four decimal places.

8. 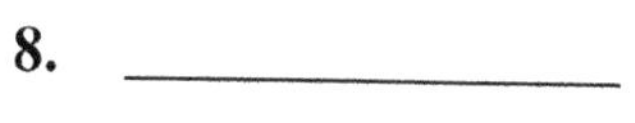

9. Graph the equation from Exercise #8 on the same coordinate axes as the data points. Does the curve appear to be a good fit for the data? Explain.

9. 

10. Determine the practical domain of the function. Round to tenths.

11. Estimate the practical range of the function. Round to nearest yard.

10. ______________

11. ______________

12. How long after the arrow was loosed did it reach 146 yards above the ground? Round to tenths.

12. ______________

Name: Date:
Instructor: Section:

13. What is the height after 1.5 seconds? Round to the nearest yard.

14. What is the height after 5.5 seconds? Round to the the nearest yard.

13. ______________

14. ______________

For #15-22, use the following scenario.

An arrow is launched with a light bow. The height, h, of the arrow above the ground in yards as a function of time, t, in seconds, can be partially modeled by the following table.

t	0	0.8	1.6	2.4	3
$h(t)$	2	35	47	39	19

15. Sketch a scatterplot of the data using your graphing calculator.

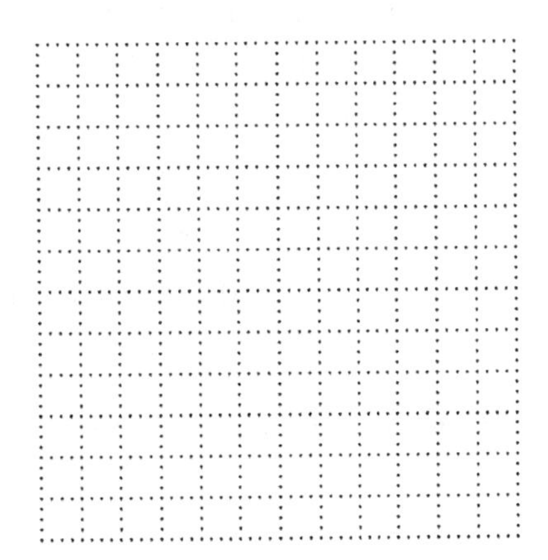

16. Use your graphing calculator to determine the quadratic regression equation for this data set. Round to four decimal places.

16. ______________

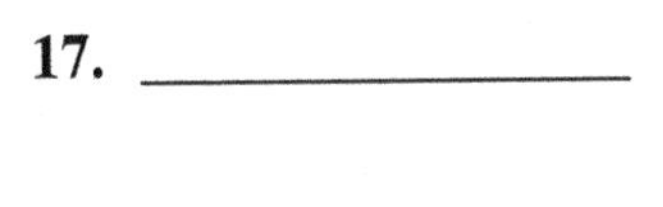

17. Graph the equation from Exercise #16 on the same coordinate axes as the data points. Does the curve appear to be a good fit for the data? Explain.

17. ______________

____

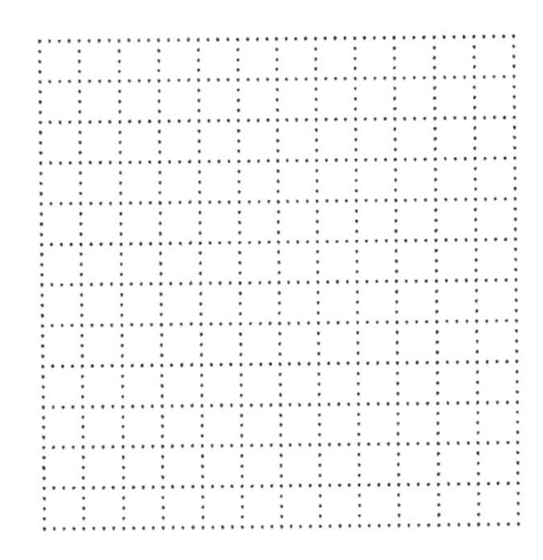

18. Determine the practical domain of the function. Round to tenths.

19. Estimate the practical range of the function. Round to nearest yard.

18. ______________

19. ______________

20. What is the height after 1 second? Round to the nearest yard.

21. What is the height after 2 seconds? Round to the nearest yard.

20. ______________

21. ______________

22. What is the height after 2.8 seconds? Round to the nearest yard.

22. ______________

For #23-28, use the following data set. Round answers to three decimal places.

x	0	5	10	15	20
y	5	35	85	200	485

23. Determine an appropriate scale, and plot these points.

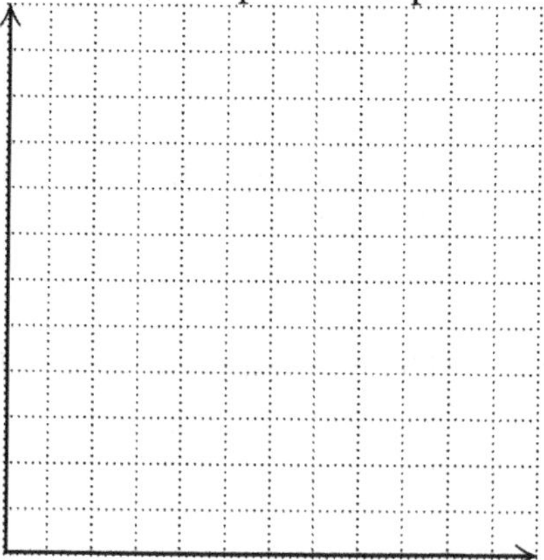

24. Use your graphing calculator to determine the quadratic regression equation for this data set.

24. ______________

Name: Date:
Instructor: Section:

25. Use your graphing calculator to graph the regression equation and data points.

26. Predict the output for $x = 8$.

27. Predict the output for $x = 12$.

26. ______________

27. ______________

28. Predict the output for $x = 17$.

28. ______________

Concept Connections

29. What model is used to make predictions about parabolic data?

____

30. Using the model from Exercise #8, the height of the arrow after 7 seconds is about –96.5 yards. Explain what is wrong with this statement.

____

____

Name: Date:
Instructor: Section:

Chapter 4 QUADRATIC AND HIGHER-ORDER POLYNOMIAL FUNCTIONS

Activity 4.7

Learning Objectives

1. Identify the imaginary unit $i = \sqrt{-1}$.
2. Identify a complex number.
3. Determine the value of the discriminant $b^2 - 4ac$.
4. Determine the types of solutions to a quadratic equation.
5. Solve a quadratic equation in the complex number system.

Key Terms

Use the vocabulary terms listed below to complete each statement in Exercises 1–4.

complex **discriminant** **imaginary** **real**

1. In the number $a + bi$, the term a is called the ________________ part.

2. Any number that can be written in the form $a + bi$ is called a ________________ number.

3. In the number $a + bi$, the term b is called the ________________ part.

4. In the quadratic equation $x = \dfrac{-b \pm \sqrt{b^2 - 4ac}}{2a}$, the expression $\sqrt{b^2 - 4ac}$ is called the ________________.

Practice Exercises

For #5-14, write each of the following in the form bi, where $i = \sqrt{-1}$.

5. $\sqrt{-144}$ **6.** $\sqrt{-28}$

5. ________________

6. ________________

7. $\sqrt{-44}$ **8.** $\sqrt{-75}$

7. ________________

8. ________________

9. $\sqrt{-27}$

10. $\sqrt{-49}$

9. ______________

10. ______________

11. $\sqrt{-32}$

12. $\sqrt{-60}$

11. ______________

12. ______________

13. $\sqrt{-\frac{49}{64}}$

14. $\sqrt{\frac{-18}{10}}$

13. ______________

14. ______________

For #15-20, perform the operations, and express your answer in the form a + bi.

15. $(3+7i)+(-5+3i)$

16. $(9-2i)-(4-8i)$

15. ______________

16. ______________

17. $2i+(9-5i)$

18. $2i(-6+5i)$

17. ______________

18. ______________

19. $(5+2i)(3-2i)$

20. $(-4+i)(2-3i)$

19. ______________

20. ______________

Name: Date:
Instructor: Section:

For #21-24, solve the quadratic equations in the complex number system using the quadratic formula.

21. $x^2 + 4x = 2$

22. $2x^2 - 3x - 9 = 0$

21. ______________

22. ______________

23. $4x^2 - x + 6 = 0$

24. $x^2 - 2x + 6 = 0$

23. ______________

24. ______________

For #25-28, determine the number and type of solutions of each equation by examining the discriminant.

25. $8x^2 + 3x + 2 = 0$

26. $3x^2 + x - 2 = 0$

25. ______________

26. ______________

27. $x^2 + 6x + 9 = 0$

28. $5x^2 - 4x + 1 = 0$

27. ______________

28. ______________

Concept Connections

29. When the discriminant of a quadratic equation is negative, what can you say about the graph of that equation?

____

____

30. Explain the number and type of solutions when the discriminant of a quadratic equation is positive, zero and negative.

____

____

Name: Date:
Instructor: Section:

Chapter 4 QUADRATIC AND HIGHER-ORDER POLYNOMIAL FUNCTIONS

Activity 4.8

Learning Objectives

1. Identify a direct variation function.
2. Determine the constant of variation.
3. Identify the properties of graphs of power functions defined by $y = kx^n$, where n is a positive integer, $k \neq 0$.

Key Terms

Use the vocabulary terms listed below to complete each statement in Exercises 1–2.

constant **direct**

1. The equation $y = kx^n$, defines a ________________ variation function.

2. In the equation $y = kx^n$, k is called the ________________ of variation.

Practice Exercises

For #3-5, assume that y varies directly as x.

3. Determine the pattern and complete the table.

x	$\frac{1}{10}$	3	6	9
y			30	

4. Determine the constant of variation.

5. Write a direct variation equation.

4. ________________

5. ________________

For #6-8, assume that y varies directly as the square of x.

6. Determine the pattern and complete the table.

x	$\frac{1}{2}$	1	4	8
y			8	

7. Determine the constant of variation.

8. Write a direct variation equation.

7. ______________

8. ______________

For #9-11, the surface area of a square is given by the function $S = 6x^2$*, where x is the side of the square.*

9. Does the surface area vary directly as the length of a side?

10. Explain your answer to Exercise #9.

9. ______________

10. ______________

____

11. What is the constant of variation?

11. ______________

For #12-15, assume that y varies directly as the cube of x.

12. Determine the direct variation function.

13. If $y = 24$, when $x = 2$, determine the constant of variation.

12. ______________

13. ______________

14. Write the particular direction variation equation.

15. Determine y when $x = 5$.

14. ______________

15. ______________

Name: Date:
Instructor: Section:

For #16-20, the distance, d that you drive varies directly as the uniform speed, s that you drive.

16. Determine the direct variation function.

17. If $d = 260$, when $s = 65$, determine the constant of variation.

16. ______________

17. ______________

18. Write the particular direction variation equation.

19. Determine d when $s = 75$.

18. ______________

19. ______________

20. Explain your answer to Exercise #19 in the context of the situation.

20. ______________

____

21. Determine the interval over which $y = -4x^4$ is increasing.

26. ______________

22. Is the graph of $y = 0.5x^6$ wider or narrower than the graph of $y = x^6$?

27. ______________

For #23-27, sketch a graph of the given power function. Verify your graphs using your graphing calculator.

23. $y = -4x^2$

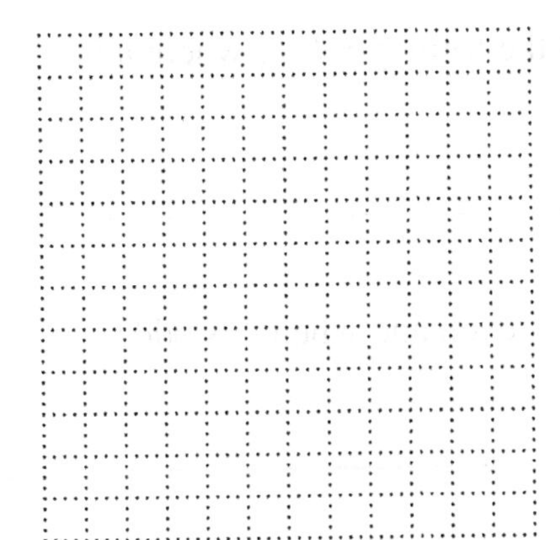

24. $y = x^4 - 1$

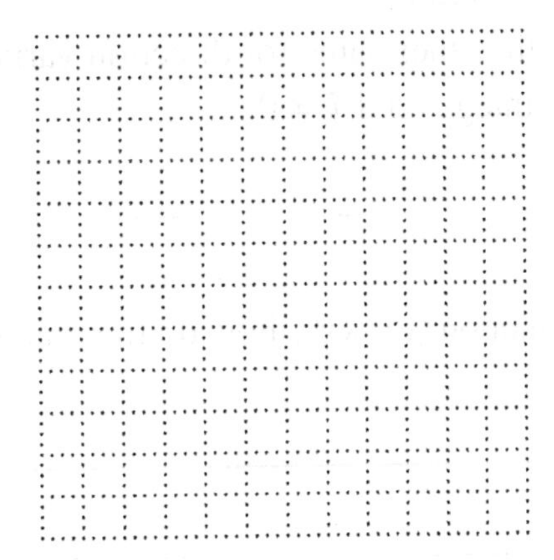

25. $y = \frac{1}{2}x^5$

26. $f(x) = x^6 + 1$

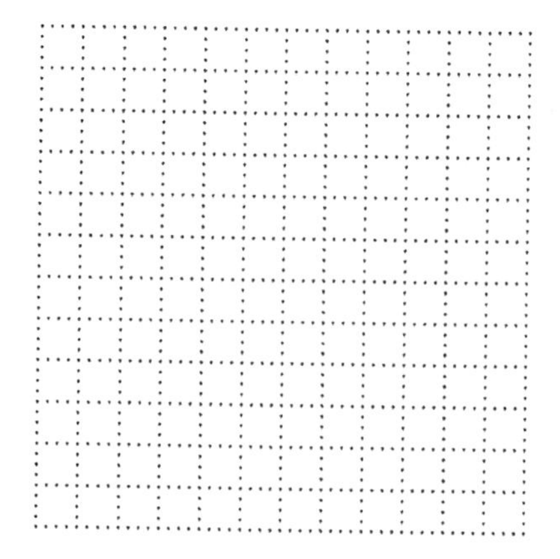

27. $g(x) = 0.1x^3 + 1$

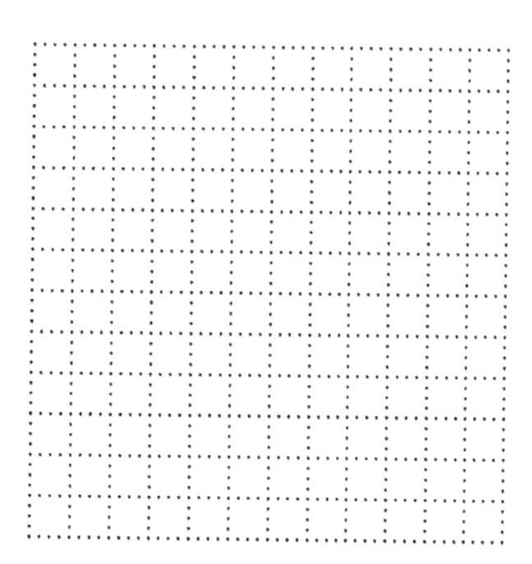

28. Does the graph $y = -5x^2$ have a maximum or a minimum point? Explain your answer.

28. ______________

____

Concept Connections

29. What is the other name for direction variation function of the form $y = kx^n$, where n is a positive integer and $k \neq 0$.

____

30. In the function $y = kx^n$, how do the graphs differ when n is even and when n is odd?

____

____

Name: Date:
Instructor: Section:

Chapter 4 QUADRATIC AND HIGHER-ORDER POLYNOMIAL FUNCTIONS

Activity 4.9

Learning Objectives

1. Identify equations that define polynomial functions.
2. Determine the degree of a polynomial function.
3. Determine the intercepts of the graph of a polynomial function.
4. Identify the properties of the graphs of polynomial functions.

Practice Exercises

For #1-2, use the function $f(x) = x^3 - x^2 - 6x$.

1. Determine the x-intercept(s) of the graph using an algebraic approach (factoring). Verify your answer using your graphing calculator.

1. ______________

2. Determine the vertical intercept.

2. ______________

For #3-4, use the function $g(x) = x^4 - 2x^3 + x^2$.

3. Determine the x-intercept(s) of the graph using an algebraic approach (factoring). Verify your answer using your graphing calculator.

3. ______________

4. Determine the vertical intercept.

4. ______________

For #5-6, use the function $h(x) = x^4 - 17x + 16$.

5. Determine the x-intercept(s) of the graph using an algebraic approach (factoring). Verify your answer using your graphing calculator.

5. 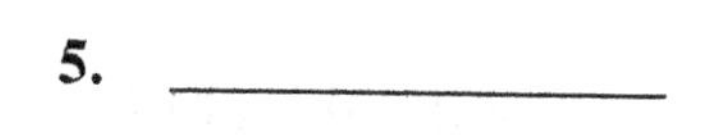

6. Determine the vertical intercept.

6. ______________

For #7-8, use the function $f(x) = x^3 - 2x^2 - x + 2$.

7. Determine the x-intercept(s) of the graph using an algebraic approach (factoring). Verify your answer using your graphing calculator.

7. ______________

8. Determine the vertical intercept.

8. ______________

For #9-10, use the function $g(x) = x^3 + 2x^2 - 25x - 50$.

9. Determine the x-intercept(s) of the graph using an algebraic approach (factoring). Verify your answer using your graphing calculator.

9. ______________

10. Determine the vertical intercept.

10. ______________

Name: Date:
Instructor: Section:

For #11-12, use the function $h(x) = x^4 + 6x^3 + 9x^2$.

11. Determine the x-intercept(s) of the graph using an algebraic approach (factoring). Verify your answer using your graphing calculator.

11. ______________

12. Determine the vertical intercept.

12. ______________

For #13-14, use the function $f(x) = x^5 - x^4 - 20x^3$.

13. Determine the x-intercept(s) of the graph using an algebraic approach (factoring). Verify your answer using your graphing calculator.

13. ______________

14. Determine the vertical intercept.

14. ______________

For #15-25, consider the polynomial function $f(x) = x^4 + 4x^3 - 5x + 4$.

15. What is the degree of the polynomial function?

16. Determine the name of this type of polynomial function.

15. ______________

16. ______________

17. Use your graphing calculator to sketch a graph of this function.

18. Determine the domain of the function.

19. Determine the range of the function.

18. ______________

19. ______________

20. Use your graphing calculator to determine the x-intercepts of the function.

21. Determine the y-intercept of the function.

20. ______________

21. ______________

22. Use your graphing calculator to determine any maximum points.

23. Use your graphing calculator to determine any minimum points.

22. ______________

23. ______________

24. Use your results to determine the interval(s) along the x-axis where the function is decreasing.

25. Use your results to determine the interval(s) along the x-axis where the function is increasing.

24. ______________

25. ______________

26. What is the degree of the polynomial function for Exercise #1?

27. What is the degree of the polynomial function for Exercise #5?

26. ______________

27. ______________

28. What is the degree of the polynomial function for Exercise #13?

28. ______________

Concept Connections

29. How do you determine the degree of a polynomial function?

____

30. Name the functions from Exercises #1, 3, 5, 7, 9, 11, and 13.

____

Name: Date:
Instructor: Section:

Chapter 4 QUADRATIC AND HIGHER-ORDER POLYNOMIAL FUNCTIONS

Activity 4.10

Learning Objectives

1. Determine the regression equation of a polynomial function that best fits the data.

Practice Exercises

For #1-10, use the following data.

x	0	5	10	15	20
$f(x)$	156.7	174	161.4	204.7	277.3

1. Plot these points on your graphing calculator. Sketch the graph.

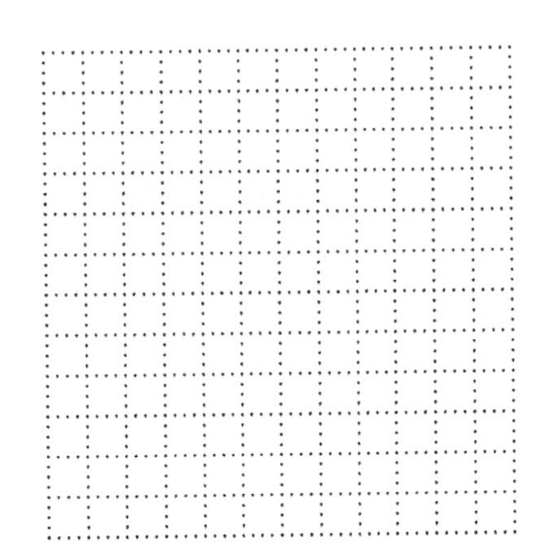

2. Using your graphing calculator, determine the regression equation of the first degree.

3. Using your graphing calculator, determine the regression equation of the second degree.

2. ______________

3. ______________

4. Using your graphing calculator, determine the regression equation of the third degree.

5. Use your calculator to fit each of your models from Exercises #2-4 to your scatterplot from #1. Which of these curves best represents the data?

4. ______________

5. ______________

6. Explain your answer to Exercise #5.

____

____

7. Use the equation from Exercise #4 to estimate $f(7.5)$. Round to the nearest unit.

8. Use the equation from Exercise #4 to estimate $f(-5)$. Round to the nearest unit.

7. ______________

8. ______________

9. Use the equation from Exercise #4 to estimate $f(25)$. Round to the nearest unit.

10. In which of these estimates (Exercises #7-9) do you have the most confidence? Explain.

9. ______________

10. ______________

____

For #11-20, use the following data.

x	0	2	4	6
$g(x)$	1.6	−4.8	97.3	20.8

11. Plot these points on your graphing calculator. Sketch the graph.

12. Using your graphing calculator, determine the regression equation of the first degree.

13. Using your graphing calculator, determine the regression equation of the second degree.

12. ______________

13. ______________

14. Using your graphing calculator, determine the regression equation of the third degree.

15. Use your calculator to fit each of your models from Exercises #12-14 to your scatterplot from #11. Which of these curves best represents the data?

14. ______________

15. ______________

Name: Date:
Instructor: Section:

16. Explain your answer to Exercise #15.

____

____

17. Use the equation from Exercise #14 to estimate $g(5)$. Round to the nearest unit.

18. Use the equation from Exercise #14 to estimate $g(10)$. Round to the nearest unit.

17. ______________

18. ______________

19. Use the equation from Exercise #14 to estimate $g(-2)$. Round to the nearest unit.

20. In which of these estimates (Exercises #17-19) do you have the most confidence? Explain.

19. ______________

20. ______________

____

For #21-28, use the following data.

x	-1	0	1	6	8
$h(x)$	0.6	1.5	-11	21	33

21. Plot these points on your graphing calculator. Sketch the graph.

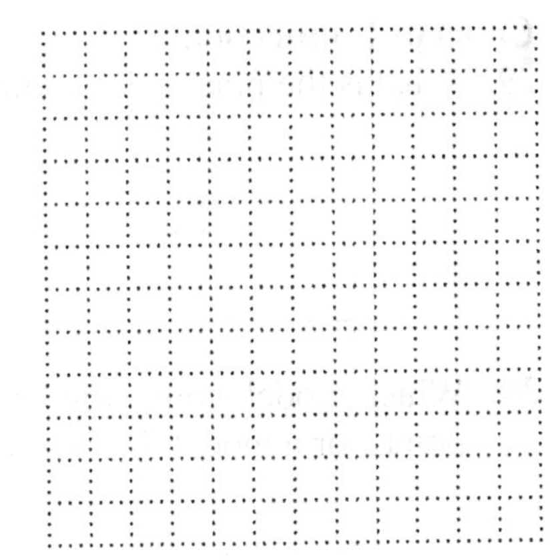

22. Using your graphing calculator, determine the regression equation of the first degree.

23. Using your graphing calculator, determine the regression equation of the second degree.

22. ______________

23. ______________

24. Using your graphing calculator, determine the regression equation of the third degree.

25. Using your graphing calculator, determine the regression equation of the fourth degree.

24. ________________

25. ________________

26. Use your calculator to fit each of your models from Exercises #22-25 to your scatterplot from #21. Which of these curves best represents the data?

26. ________________

27. Explain your answer to Exercise #26.

____

____

28. Use the equation from Exercise #25 to estimate $h(5)$. Round to the nearest unit.

28. ________________

Concept Connections

29. What is the practical domain and range of the data from Exercise #1?

____

____

30. Which model would you be more confident in: a model that was created from 6 data points, or a model that was created from 10 data points? Explain.

____

____

Name: Date:
Instructor: Section:

Chapter 5 RATIONAL AND RADICAL FUNCTIONS

Activity 5.1

Learning Objectives

1. Determine the domain and range of a function defined by $y = \frac{k}{x}$, where k is a nonzero real number.
2. Determine the vertical and horizontal asymptotes of the graph of $y = \frac{k}{x}$.
3. Sketch a graph of functions of the form $y = \frac{k}{x}$.
4. Determine the properties of graphs having equation $y = \frac{k}{x}$.

Practice Exercises

For #1-12, consider the function $f(x) = \frac{3}{x}$.

1. What is the domain of function f? **1.** ______________

2. Complete the table.

x	-10	-5	-1	-0.1	0	0.1	1	5	10
$f(x)$									

3. Sketch the graph of this function.

4. Using the table and the graph, determine what happens to the y-values as the x-values increase infinitely in the positive direction.

5. Using the table and the graph, determine what happens to the y-values as the x-values decrease infinitely in the negative direction.

6. What is the horizontal asymptote for the graph?

7. What happens to the y-values as the positive x-values get close to zero?

8. What happens to the y-values as the negative x-values get close to zero?

9. What is the vertical asymptote for the graph?

10. Does the function have a maximum or minimum functional value?

11. Is the function continuous?

12. Determine the symmetry, if any.

4. ______________

____

5. ______________

____

6. ______________

7. ______________

____

8. ______________

____

9. ______________

10. ______________

11. ______________

____

12. ______________

____

Name: Date:
Instructor: Section:

For #13-23, consider the function $g(x) = -\frac{3}{x}$.

13. What is the domain of function g? **13.** ______________

14. Complete the table.

x	-10	-5	-1	-0.1	0	0.1	1	5	10
$g(x)$									

15. Sketch the graph of this function.

16. Using the table and the graph, determine what happens to the y-values as the x-values increase infinitely in the positive direction.

17. Using the table and the graph, determine what happens to the y-values as the x-values decrease infinitely in the negative direction.

16. ______________

____

17. ______________

____

18. What is the horizontal asymptote for the graph?

19. What happens to the y-values as the positive x-values get close to zero?

18. ______________

19. ______________

____

20. What happens to the y-values as the negative x-values get close to zero?

21. What is the vertical asymptote for the graph?

20. ______________

____

21. ______________

22. Does the function have a maximum or minimum functional value?

23. Describe how to obtain the graph of the function g using the graph of the function f.

22. ______________

23. ______________

____

For #24-28, consider a trip of 600 miles.

24. Write an equation that defines t as a function of r, in which r represents the average speed in miles per hour and t represents the time in hours to complete the trip.

24. ______________

25. Complete the table.

Input, r	50	55	60	65	70	75
Output, t						

26. As your average speed increases, what happens to the time it takes to complete the trip?

27. As your average speed for the trip gets closer to zero, what happens to the time it takes to complete the 600-mile trip?

26. ______________

____

27. ______________

____

28. What is the practical domain?

28. ______________

Concept Connections

29. Define horizontal asymptote.

____

____

30. Define vertical asymptote.

____

____

Name: Date:
Instructor: Section:

Chapter 5 RATIONAL AND RADICAL FUNCTIONS

Activity 5.2

Learning Objectives

1. Graph a function defined by an equation of the form $y = \frac{k}{x^n}$ where n is any positive integer and k is a nonzero real number, $x \neq 0$.
2. Describe the properties of graphs having equation $y = \frac{k}{x^n}$, $x \neq 0$.
3. Determine the constant of proportionality (also called the constant of variation).

Practice Exercises

For #1-5, let y vary inversely as the square of x.

1. Determine the equation for y.

2. If $y = 8$ when $x = 3$, determine the constant of proportionality.

1. ______________

2. ______________

3. Write the particular equation using this constant.

4. If $x = 4$, determine y.

3. ______________

4. ______________

5. If the value of x is multiplied by 10, what is the effect on the value of y?

5. ______________

For #6-17, consider the function $f(x) = \frac{1}{x^3}$.

6. What is the domain of the function f?

6. ______________

7. Complete the table.

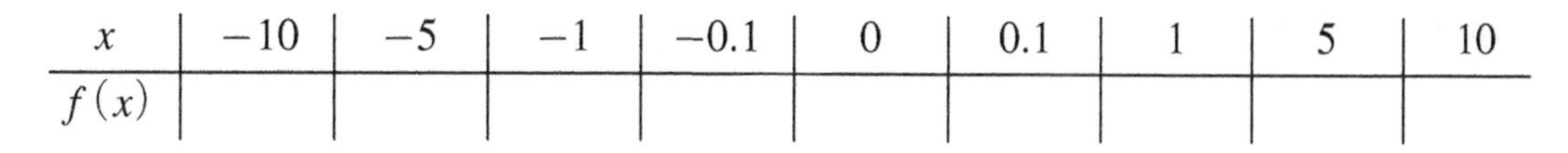

x	-10	-5	-1	-0.1	0	0.1	1	5	10
$f(x)$									

8. Use your graphing calculator to make a graph. Sketch the graph below.

9. Using the table and the graph, determine what happens to the y-values as the x-values increase infinitely in the positive direction.

10. Using the table and the graph, determine what happens to the y-values as the x-values decrease infinitely in the negative direction.

9. ______________

____

10. ______________

____

11. What is the horizontal asymptote for the graph?

12. What happens to the y-values as the positive x-values get close to zero?

11. ______________

12. ______________

____

13. What happens to the y-values as the negative x-values get close to zero?

14. What is the vertical asymptote for the graph?

13. ______________

____

14. ______________

15. Does the function have a maximum or minimum functional value?

16. Is the function continuous?

15. ______________

16. ______________

____

17. Determine the symmetry, if any.

17. ______________

Name: Date:
Instructor: Section:

For #18-28, consider the function $g(x) = -\frac{1}{x^3}$.

18. What is the domain of the function g ?

18. ______________

19. Complete the table.

x	-10	-5	-1	-0.1	0	0.1	1	5	10
$g(x)$									

20. Use your graphing calculator to make a graph. Sketch the graph below.

21. Using the table and the graph, determine what happens to the y-values as the x-values increase infinitely in the positive direction.

22. Using the table and the graph, determine what happens to the y-values as the x-values decrease infinitely in the negative direction.

21. ______________

____

22. ______________

____

23. What is the horizontal asymptote for the graph?

24. What happens to the y-values as the positive x-values get close to zero?

23. ______________

24. ______________

____

25. What happens to the y-values as the negative x-values get close to zero?

26. What is the vertical asymptote for the graph?

25. ______________

____

26. ______________

27. Does the function have a maximum or minimum functional value?

28. Describe how to obtain the graph of the function g using the graph of the function f.

27. ______________

28. ______________

____

Concept Connections

29. Give another name for the constant of proportionality and explain how it is used in an inverse variation function.

____

____

30. If y varies inversely as the cube of x and $y = \frac{3}{4}$ when $x = 8$, determine the equation for y.

____

____

Name: Date:
Instructor: Section:

Chapter 5 RATIONAL AND RADICAL FUNCTIONS

Activity 5.3

Learning Objectives

1. Determine the domain of a rational function defined by an equation of the form $y=\frac{k}{g(x)}$ where k is a nonzero constant and $g(x)$ is a first-degree polynomial.
2. Identify the vertical and horizontal asymptotes of $y=\frac{k}{g(x)}$.
3. Sketch graphs of rational functions defined by $y=\frac{k}{g(x)}$.

Practice Exercises

For #1-12, consider the function $g(x)=\frac{4}{x+4}$.

1. What is the domain of the function g? **1.** ________________

2. Complete the following table.

x	-8	-5	-4.1	-4.01	-4	-3.99	-3.9	-3	0
$g(x)$									

3. Use your graphing calculator to sketch the graph of this function.

4. Does the graph have a horizontal asymptote?

5. Explain your answer to Exercise #4.

4. ________________

5. ________________

____

6. What is the equation of the horizontal asymptote?

7. Does the graph have a vertical asymptote?

6. ______________

7. ______________

8. Explain your answer to Exercise #7.

9. What is the equation of the vertical asymptote?

8. ______________

____

9. ______________

10. For what value of x is $g(x)$ a maximum?

11. Determine the y-intercept.

10. ______________

11. ______________

12. Determine the x-intercept.

12. ______________

For #13-19, consider the function $f(x) = \dfrac{4}{x-13}$.

13. What is the domain of the function f?

13. ______________

14. Determine the vertical asymptote of f.

15. Explain your answer to Exercise #14.

14. ______________

15. ______________

____

16. Determine the horizontal asymptote.

17. Determine the y-intercept.

16. ______________

17. ______________

Name: Date:
Instructor: Section:

18. Determine the x-intercept.

19. For what value of x is $f(x)$ a maximum?

18. ______________

19. ______________

For #20-24, consider the function $h(x) = \dfrac{-5}{12-x}$.

20. What is the domain of function h?

21. Determine the vertical asymptote of h.

20. ______________

21. ______________

22. Use your graphing calculator to sketch the graph of this function.

23. Determine the horizontal asymptote.

24. Determine the y-intercept.

23. ______________

24. ______________

For #25-28, consider the function $F(x)=\dfrac{22}{1.5x-30}$.

25. What is the domain of function F?

26. Determine the vertical asymptote of F.

25. ______________

26. ______________

27. Use your graphing calculator to sketch the graph of this function.

28. Determine the horizontal asymptote.

28. ______________

Concept Connections

29. Describe how to find the vertical asymptote of a rational function in general form, $Q(x)=\dfrac{k}{g(x)}$.

____

____

30. Describe how to find the horizontal asymptote of a rational function in general form, $Q(x)=\dfrac{k}{g(x)}$.

____

____

Name: Date:
Instructor: Section:

Chapter 5 RATIONAL AND RADICAL FUNCTIONS

Activity 5.4

Learning Objectives

1. Solve an equation involving a rational expression using an algebraic approach.
2. Solve an equation involving a rational expression using a graphing approach.
3. Determine horizontal asymptotes of the graph of $y = \frac{f(x)}{g(x)}$, where $f(x)$ and $g(x)$ are first-degree polynomials.

Practice Exercises

For #1-12, solve each equation algebraically and check by graphing.

1. $\frac{2x}{2x-3} = 4$

2. $\frac{3x}{x-2} = 1$

3. $\frac{x+6}{2x+3} = \frac{3}{2}$

4. $\frac{x-1}{x+3} = -\frac{1}{5}$

5. $\frac{x-4}{3x+5} = \frac{1}{20}$

6. $\frac{x+9}{x-7} = -3$

1. ______________

2. ______________

3. ______________

4. ______________

5. ______________

6. ______________

7. $\dfrac{5-x}{1-x}=2$

8. $\dfrac{3-x}{3x+1}=-1$

7. ________________

8. ________________

9. $\dfrac{5x+4}{x+2}=2$

10. $\dfrac{-3x}{2-x}=5$

9. ________________

10. ________________

11. $\dfrac{x-2}{3x-4}=3$

12. $\dfrac{x+6}{x+3}=6$

11. ________________

12. ________________

For #13-20, determine the domain and the vertical asymptotes of each rational function.

13. $y=\dfrac{3x}{x-2}$

14. $y=\dfrac{x-1}{x+3}$

13. ________________

____

14. ________________

____

15. $y=\dfrac{x+9}{x-7}$

16. $y=\dfrac{3-x}{3x+1}$

15. ________________

____

16. ________________

____

Name: Date:
Instructor: Section:

17. $y = \dfrac{x-2}{3x-4}$

18. $y = \dfrac{5x+4}{x+2}$

19. $y = \dfrac{x+6}{2x+3}$

20. $y = \dfrac{2x}{2x-3}$

17. ______________
____

18. ______________
____

19. ______________
____

20. ______________
____

For #21-28, determine the horizontal asymptote of each rational function by inspecting the graph of the function using your graphing calculator.

21. $y = \dfrac{3x}{x-2}$

22. $y = \dfrac{x-1}{x+3}$

23. $y = \dfrac{x+9}{x-7}$

24. $y = \dfrac{3-x}{3x+1}$

25. $y = \dfrac{-3x}{2-x}$

26. $y = \dfrac{5x+4}{x+2}$

27. $y = \dfrac{x+6}{2x+3}$

28. $y = \dfrac{2x}{2x-3}$

21. ______________

22. ______________

23. ______________

24. ______________

25. ______________

26. ______________

27. ______________

28. ______________

Concept Connections

29. Solve $\frac{7x-9}{x+5}=4$ by graphing using your graphing calculator. Round to the nearest tenth.

---

30. Solve $\frac{x+3}{x-3}=1$ by graphing using your graphing calculator. Sketch the graph. Why can't your calculator solve this equation?

Name: Date:
Instructor: Section:

Chapter 5 RATIONAL AND RADICAL FUNCTIONS

Activity 5.5

Learning Objectives

1. Determine the least common denominator (LCD) of two or more rational expressions.
2. Solve an equation involving rational expressions using an algebraic approach.
3. Solve a formula for a specific variable.

Practice Exercises

For #1-10, find the LCD.

1. $\frac{3}{7}$ and $\frac{5}{6}$

2. $\frac{2}{9}$ and $\frac{8}{11}$

3. $\frac{8}{15}$ and $\frac{7}{27}$

4. $\frac{5}{18}$ and $\frac{11}{12}$

5. $\frac{9}{4x^2}$ and $\frac{3}{8x^5}$

6. $\frac{7}{10y^3}$ and $\frac{8}{25y^4}$

7. $\frac{13y}{14x^2z^3}$ and $\frac{25x^3}{48y^2z}$

8. $\frac{3}{20x^6y^2z^3}$ and $\frac{5}{12xyz^4}$

9. $\frac{6}{x-2}$ and $\frac{2}{x+1}$

10. $\frac{x}{x+3}$ and $\frac{3x}{x-1}$

1. ____________

2. ____________

3. ____________

4. ____________

5. ____________

6. ____________

7. ____________

8. ____________

9. ____________

10. ____________

For #11-24, solve each equation using an algebraic approach. Round to 2 decimal places.

11. $\frac{3x}{7}+\frac{5x}{6}=\frac{1}{14}$

12. $\frac{2x}{9}+\frac{8x}{11}=47$

11. ______________

12. ______________

13. $\frac{1}{5x}+\frac{1}{3x}=1$

14. $\frac{1}{18x}+\frac{1}{12x}=\frac{1}{6}$

13. ______________

14. ______________

15. $\frac{4}{x}+\frac{6}{x}=1$

16. $\frac{1}{10x}+\frac{1}{25x}=1$

15. ______________

16. ______________

17. $\frac{6}{x-2}=8$

18. $\frac{2}{x+1}=5$

17. ______________

18. ______________

19. $\frac{7}{x+3}=2$

20. $\frac{3}{x-1}=4$

19. ______________

20. ______________

Name: Date:
Instructor: Section:

21. $\frac{1}{t}+\frac{1}{t+3}=\frac{1}{3}$

22. $\frac{1}{t}+\frac{1}{t+4}=\frac{1}{4}$

21. ______________

22. ______________

23. $\frac{1}{4}+\frac{1}{6}=\frac{1}{x}$

24. $\frac{1}{9}+\frac{1}{6}=\frac{1}{x}$

23. ______________

24. ______________

For #25-28, solve each equation for the indicated variable.

25. $\frac{2}{x}+\frac{3}{y}=\frac{1}{z}$ for z

26. $\frac{2}{x}+\frac{3}{y}=\frac{1}{z}$ for y

25. ______________

26. ______________

27. $\frac{4}{a+b}=\frac{5}{c}$ for c

28. $\frac{4}{a+b}=\frac{5}{c}$ for a

27. ______________

28. ______________

Concept Connections

29. Tom can mow his lawn 10 minutes faster than Bill. If working together they can mow the lawn in 30 minutes, how long does it take each of them working alone? Use the formula $\frac{1}{t_1}+\frac{1}{t_2}=\frac{1}{T}$. Round to the nearest minute.

30. Three pumps work together to fill a tank, with filling times of 50 minutes, 60 minutes and 75 minutes. How long will it take to fill the tank if all three pumps are working simultaneously? Use the formula $\frac{1}{t_1}+\frac{1}{t_2}+\frac{1}{t_3}=\frac{1}{T}$.

Name: Date:
Instructor: Section:

Chapter 5 RATIONAL AND RADICAL FUNCTIONS

Activity 5.6

Learning Objectives
1. Multiply and divide rational expressions.
2. Add and subtract rational expressions.
3. Simplify a complex fraction.

Practice Exercises

For #1-4, multiply and simplify the rational expressions.

1. $\frac{8x}{9y} \cdot \frac{3y}{4x}$

2. $\frac{20a^3}{21b} \cdot \frac{7b^4}{40a^4}$

3. $\frac{8(x+1)}{12(x-2)} \cdot \frac{9(x+2)}{2(x+1)}$

4. $\frac{x^2-3x+2}{x^2-1} \cdot \frac{x^2-2x-3}{x^2-9}$

1. ______________

2. ______________

3. ______________

4. ______________

For #5-8, divide and simplify the rational expressions.

5. $\frac{5a^2}{2b^3} \div \frac{35a}{2b}$

6. $\frac{36x^3}{35y^2} \div \frac{18x^4}{21y^5}$

7. $\frac{10(x-4)}{3(x+1)} \div \frac{14(x+4)}{9(x+1)}$

8. $\frac{x^2-4}{x^2+x-2} \div \frac{x^2+3x-10}{x^2-6x+5}$

5. ______________

6. ______________

7. ______________

8. ______________

For #9-12, add and simplify the rational expressions.

9. $3+\frac{5}{x}$

10. $\frac{4}{x}+\frac{7}{x}$

9. ______________

10. ______________

11. $\frac{x+1}{3}+\frac{x-1}{2}$

12. $\frac{5x}{3y}+\frac{4y}{7x}$

11. ______________

12. ______________

13. $\frac{x+3}{x+2}+\frac{x+5}{x-3}$

14. $\frac{8}{x^2-x}+\frac{11}{x^2+x}$

13. ______________

14. ______________

For #15-20, subtract and simplify the rational expressions.

15. $\frac{2}{x}-5$

16. $\frac{6}{x}-\frac{7}{x^2}$

15. ______________

16. ______________

17. $\frac{x+2}{4}-\frac{x-3}{5}$

18. $\frac{8y}{5x}-\frac{8x}{3y}$

17. ______________

18. ______________

19. $\frac{x-1}{x-3}-\frac{x+2}{x+4}$

20. $\frac{2}{x^2-3x}-\frac{4}{x^2+3x}$

19. ______________

20. ______________

Name: Date:
Instructor: Section:

For #21-28, simplify each complex fraction.

21. $\dfrac{\dfrac{100}{x}}{\dfrac{300}{x^2+3x}}$

22. $\dfrac{\dfrac{30}{x^2-5x}}{\dfrac{24}{x}}$

21. ________________

22. ________________

23. $\dfrac{3+\dfrac{5}{x}}{\dfrac{25}{x^2}+\dfrac{15}{x}}$

24. $\dfrac{\dfrac{3}{x}+2}{\dfrac{4}{x}+\dfrac{6}{x^2}}$

23. ________________

24. ________________

25. $\dfrac{\dfrac{3}{x+1}+\dfrac{5}{x}}{\dfrac{5}{x+1}+\dfrac{2}{x}}$

26. $\dfrac{\dfrac{1}{x+2}-\dfrac{3}{x-1}}{1+\dfrac{x+8}{x-1}}$

25. ________________

26. ________________

27. $\dfrac{x-\dfrac{49}{x}}{\dfrac{7}{x}-1}$

28. $\dfrac{\dfrac{x}{5}-1}{x-\dfrac{25}{x}}$

27. ________________

28. ________________

Concept Connections

29. How is a complex fraction different from any other fraction?

30. Explain how to simplify a complex fraction by simplifying the numerator and denominator.

Name: Date:
Instructor: Section:

Chapter 5 RATIONAL AND RADICAL FUNCTIONS

Activity 5.7

Learning Objectives

1. Determine the domain of a radical function defined by $y = \sqrt{g(x)}$, where $g(x)$ is a polynomial.
2. Graph functions having equation $y = \sqrt{g(x)}$ and $y = -\sqrt{g(x)}$.
3. Identify the properties of the graph of $y = \sqrt{g(x)}$ and $y = -\sqrt{g(x)}$.

Key Terms

Use the vocabulary terms listed below to complete each statement in Exercises 1–3.

index **radical sign** **radicand**

1. The expression under the radical sign is called the ____________________ .

2. The 2 in the expression $\sqrt[2]{n}$ is called the ____________________ .

3. The symbol $\sqrt{\ }$ is called the ____________________ .

Practice Exercises

For #4-11, using your calculator, determine the value of each number to the nearest hundredth, if necessary.

4. $\sqrt{42}$ **5.** $\sqrt{50}$

4. ________________

5. ________________

6. $10^{1/2}$ **7.** $21^{1/2}$

6. ________________

7. ________________

8. $\left(\sqrt{15}\right)^4$

9. $\left(\sqrt{8}\right)^4$

8. ______________

9. ______________

10. $\left(25^{1/2}\right)^3$

11. $\left(49^{1/2}\right)^3$

10. ______________

11. ______________

For #12-19, determine the domain of each function.

12. $f(x)=\sqrt{x+9}$

13. $g(x)=\sqrt{x-6}$

12. ______________

13. ______________

14. $h(x)=\sqrt{2x+1}$

15. $k(x)=\sqrt{5x+3}$

14. ______________

15. ______________

16. $R(x)=\sqrt{8-4x}$

17. $Q(x)=\sqrt{10-2x}$

16. ______________

17. ______________

18. $u(x)=-\sqrt{4x}$

19. $v(x)=3-\sqrt{5x}$

18. ______________

19. ______________

Name: Date:
Instructor: Section:

For #20-23, use the function $f(x)=\sqrt{x+5}$.

20. Determine the domain of the function.

21. Determine the x-intercept of the function.

20. ________________

21. ________________

22. Determine the y-intercept of the function.

22. ________________

23. Sketch the graph of this function.

For #24-26, use the function $g(x)=\sqrt{x}+5$.

24. Determine the domain of the function.

25. Determine the y-intercept of the function.

24. 

25. ________________

26. Sketch the graph of this function.

For #27-28, use the function $g(x)=\sqrt{x}-2$.

27. Determine the x- and y-intercepts of the function.

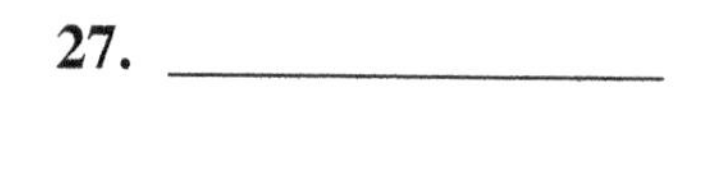

28. Sketch the graph of this function.

Concept Connections

29. What is the relationship between the graphs of $y=-\sqrt{f(x)}$ and $y=\sqrt{f(x)}$?

---

30. Is $f(x)=\sqrt{5}+x$ a radical function? What is the domain of $f(x)=\sqrt{5}+x$?

Name: Date:
Instructor: Section:

Chapter 5 RATIONAL AND RADICAL FUNCTIONS

Activity 5.8

Learning Objectives

1. Solve an equation involving a radical expression using a graphical and algebraic approach.

Practice Exercises

For #1-22, solve each equation algebraically and check by graphing. Be aware of any extraneous roots.

1. $\sqrt{x} = 5.8$

2. $\sqrt{x} = 14.2$

3. $\sqrt{x} - 6 = 0$

4. $\sqrt{x} - 10 = 0$

5. $\sqrt{3x} = 9$

6. $\sqrt{6x} = 7$

7. $3\sqrt{x} = 21$

8. $5\sqrt{x} = 20$

9. $6\sqrt{2x} = 18$

10. $8\sqrt{5x} = 80$

1. ______________

2. ______________

3. ______________

4. ______________

5. ______________

6. ______________

7. ______________

8. ______________

9. ______________

10. ______________

11. $\sqrt{x+8}=2$

12. $\sqrt{x+5}=6.9$

13. $\sqrt{x-3}=4$

14. $\sqrt{x-9}=12$

15. $\sqrt{x+1}+8=0$

16. $\sqrt{x+3}+9=0$

17. $\sqrt{x-2}+3=5$

18. $\sqrt{x-7}+1=4$

19. $\sqrt{x+14}-3=0$

20. $\sqrt{x+11}-2=0$

21. $\sqrt{x+2}=-x$

22. $\sqrt{3x+4}=-x$

11. ______________

12. ______________

13. ______________

14. ______________

15. ______________

16. ______________

17. ______________

18. ______________

19. ______________

20. ______________

21. ______________

22. ______________

Name: Date:
Instructor: Section:

For #23-26, solve each equation algebraically and by graphing. Round to the nearest hundredth.

23. $\sqrt{3x-6}=2.1$

24. $9\sqrt{x+4}=8$

25. $\sqrt{1-x}=x+2$

26. $\sqrt{3.7x+19.8}=\sqrt{2-4.5x}$

23. ______________

24. ______________

25. ______________

26. ______________

For #27-28, the time, t, in seconds, that it takes for a pendulum to complete one complete period (to swing back and forth one time) is modeled by $t=2\pi\sqrt{\frac{L}{32}}$ *where L is the length of the pendulum, in feet. Round to the nearest tenth.*

27. How long is the pendulum of a clock with a period of 1.5 seconds?

28. How long is the pendulum of a clock with a period of 2.5 seconds?

27. ______________

28. ______________

Concept Connections

29. What is a radical equation?

____

____

30. What is an extraneous solution?

____

____

Name: Date:
Instructor: Section:

Chapter 5 RATIONAL AND RADICAL FUNCTIONS

Activity 5.9

Learning Objectives

1. Determine the domain of a function defined by an equation of the form $y = \sqrt[n]{g(x)}$ where n is a positive integer and $g(x)$ is a polynomial.
2. Graph $y = \sqrt[n]{g(x)}$.
3. Identify the properties of graphs of $y = \sqrt[n]{g(x)}$.
4. Solve radical equations that contain radical expressions with an index other than 2.

Practice Exercises

For #1-8, if possible, determine the exact value of each of the following.

1. $\sqrt[3]{125}$

2. $\sqrt[7]{128}$

3. $(-216)^{1/3}$

4. $(256)^{1/4}$

5. $\sqrt{\frac{1}{144}}$

6. $(-10{,}000)^{1/4}$

7. $(32)^{1/5}$

8. $(-1)^{1/8}$

1. ____________

2. ____________

3. ____________

4. ____________

5. ____________

6. ____________

7. ____________

8. ____________

For #9-16, determine the domain of each function.

9. $y=\sqrt[3]{x-11}$

10. $y=\sqrt[3]{1.5-x}$

9. ________________

10. ________________

11. $y=\sqrt[4]{4-x}$

12. $y=\sqrt[4]{x-9}$

11. ________________

12. ________________

13. $y=\sqrt[5]{3.2-x}$

14. $y=\sqrt[5]{49+x}$

13. ________________

14. ________________

15. $y=\sqrt[6]{x+8.6}$

16. $y=\sqrt[6]{7-x}$

15. ________________

16. ________________

For #17-28, solve each equation algebraically and verify your results graphically.

17. $\sqrt[3]{x-6}=4$

18. $\sqrt[3]{x+9}=1$

17. ________________

18. ________________

19. $\sqrt[3]{7-x}=3$

20. $\sqrt[3]{4+x}=2$

19. ________________

20. ________________

Name: Date:
Instructor: Section:

21. $\sqrt[4]{4-x}=5$

22. $\sqrt[4]{x-9}=3$

21. ______________

22. ______________

23. $\sqrt[4]{3.2-x}+2=6$

24. $\sqrt[4]{x+8}+7=9$

23. ______________

24. ______________

25. $\sqrt[3]{x+5}+4=2$

26. $\sqrt[3]{2-x}=-4$

25. ______________

26. ______________

27. $x^{2/3}=49$

28. $5x^{3/4}=320$

27. ______________

28. ______________

Concept Connections

29. How is the domain of the function $f(x)=\sqrt[3]{x}$ different from the domain of $g(x)=\sqrt[4]{x}$?

____

____

30. The function for the basal metabolic rate (BMR) is $B(w)=70\sqrt[4]{w^3}$. Solve for w.

____

Name: Date:
Instructor: Section:

Chapter 6 INTRODUCTION TO THE TRIGONOMETRIC FUNCTIONS

Activity 6.1

Learning Objectives

1. Identify the sides and corresponding angles of a right triangle.
2. Determine the length of the sides of similar right triangles using proportions.
3. Determine the sine, cosine, and tangent of an angle using a right triangle.
4. Determine the sine, cosine, and tangent of an acute angle by using the graphing calculator.

Key Terms

Use the vocabulary terms listed below to complete each statement in Exercises 1–2.

similar **acute**

1. An angle A which measures $0° \le A < 90°$ is called a(n) ________________ angle.

2. Triangles whose angles are the same but whose sides are different lengths are called ________________ triangles.

Practice Exercises

For #3-16, solve the following equations. Round your answer to the nearest tenth.

3. $\cos 32° = \dfrac{x}{11}$

4. $\tan 74° = \dfrac{x}{21}$

5. $\sin 55° = \dfrac{x}{8}$

6. $\cos 41° = \dfrac{12}{x}$

7. $\tan 16° = \dfrac{32}{x}$

8. $\sin 29° = \dfrac{5}{x}$

3. ________________

4. ________________

5. ________________

6. ________________

7. ________________

8. ________________

9. $\sin 38^\circ = \frac{x}{17}$ **10.** $\cos 15^\circ = \frac{93}{x}$

9. ______________

10. ______________

11. $\tan 67^\circ = \frac{29}{x}$ **12.** $\tan 4^\circ = \frac{53}{x}$

11. ______________

12. ______________

13. $\cos 88^\circ = \frac{41}{x}$ **14.** $\sin 79^\circ = \frac{x}{25}$

13. ______________

14. ______________

15. $\sin 49^\circ = \frac{6}{x}$ **16.** $\cos 25^\circ = \frac{x}{15}$

15. ______________

16. ______________

For #17-22, triangle ABC is a right triangle. Determine each of the following.

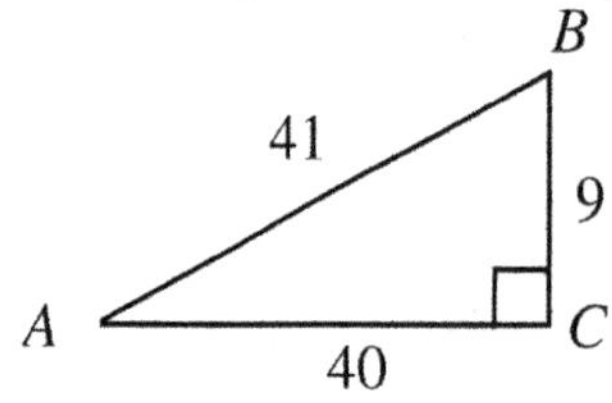

17. $\cos A$ **18.** $\cos B$

17. ______________

18. ______________

19. $\sin A$ **20.** $\sin B$

19. ______________

20. ______________

Name: Date:
Instructor: Section:

21. $\tan B$

22. $\tan A$

21. ______________

22. ______________

For #23-25, the length of a shorter leg of a 30-60-90 right triangle is 25 ft.

23. Sketch and label the diagram.

24. Determine the length of the longer leg.

25. Determine the length of the hypotenuse.

24. ______________

25. ______________

For #26-28, the length of the hypotenuse of a 30-60-90 right triangle is 15 m.

26. Sketch and label the diagram.

27. Determine the length of the shorter leg.

28. Determine the length of the longer leg.

27. ______________

28. ______________

Concept Connections

29. Draw a right triangle with acute angle θ, labeling the hypotenuse, adjacent, and opposite sides with respect to θ.

30. What is the purpose of SOH CAH TOA, and what does it stand for?

____

____

____

Name: Date:
Instructor: Section:

Chapter 6 INTRODUCTION TO THE TRIGONOMETRIC FUNCTIONS

Activity 6.2

Learning Objectives

1. Identify complementary angles.
2. Demonstrate that the sine of one of the complementary angles equals the cosine of the other.

Practice Exercises

For #1-11, find the angle complementary to the given angle.

1. 8° **2.** 18°

3. 25° **4.** 30°

5. 37° **6.** 42°

7. 50° **8.** 55°

9. 63° **10.** 71°

11. 83°

1. ________
2. ________
3. ________
4. ________
5. ________
6. ________
7. ________
8. ________
9. ________
10. ________
11. ________

For #12-15, find the sine of the given angle to four decimal places.

12. 8° **13.** 18° **12.** ____________

13. ____________

14. 25° **15.** 30° **14.** ____________

15. ____________

For #16-19, find the cosine of the given angle to four decimal places.

16. 37° **17.** 42° **16.** ____________

17. ____________

18. 50° **19.** 55° **18.** ____________

19. ____________

For #20-23, find the sine of the complement of the given angle to four decimal places.

20. 50° **21.** 55° **20.** ____________

21. ____________

22. 63° **23.** 71° **22.** ____________

23. ____________

Name: Date:
Instructor: Section:

For #24-28, find the cosine of the complement of the given angle to four decimal places.

24. 25°

25. 30°

26. 37°

27. 42°

28. 83°

24. ______________

25. ______________

26. ______________

27. ______________

28. ______________

Concept Connections

29. Give an example of an acute angle x where $\cos x = \sin(90° - x)$.

____

____

30. For a 45°-45°-90° triangle, how do you find the exact values for $\cos x$ and $\sin x$?

____

____

Name: Date:
Instructor: Section:

Chapter 6 INTRODUCTION TO THE TRIGONOMETRIC FUNCTIONS

Activity 6.3

Learning Objectives

1. Determine the inverse tangent of a number.
2. Determine the inverse sine and cosine of a number using the graphing calculator.
3. Identify the domain and range of the inverse sine, cosine, and tangent functions.

Practice Exercises

For #1-20, use your calculator to determine θ to the nearest 0.01°.

1. $\theta = \arctan 1$

2. $\theta = \sin^{-1}\left(\frac{\sqrt{3}}{2}\right)$

3. $\theta = \arccos\left(\frac{3}{8}\right)$

4. $\theta = \tan^{-1}(3.7629)$

5. $\theta = \arcsin(0.4045)$

6. $\theta = \cos^{-1}(0.6188)$

7. $\theta = \sin^{-1}(0.3426)$

8. $\theta = \arctan(0.9583)$

9. $\theta = \arccos\left(\frac{4}{9}\right)$

10. $\theta = \sin^{-1}\left(\frac{4}{9}\right)$

1. ________________

2. ________________

3. ________________

4. ________________

5. ________________

6. ________________

7. ________________

8. ________________

9. ________________

10. ________________

11. $\sin\theta = \frac{3}{8}$

12. $\tan\theta = \frac{8}{3}$

11. ______________

12. ______________

13. $\cos\theta = \frac{3}{8}$

14. $\tan\theta = \frac{3}{8}$

13. ______________

14. ______________

15. $\cos\theta = 0.5714$

16. $\tan\theta = 0.5714$

15. ______________

16. ______________

17. $\sin\theta = 0.5714$

18. $\sin\theta = 0.2689$

17. ______________

18. ______________

19. $\tan\theta = 0.2689$

20. $\cos\theta = 0.2689$

19. ______________

20. ______________

For #21-25, for each of the following, determine θ without using your calculator.

21. $\tan\theta = \frac{1}{\sqrt{3}}$

22. $\sin\theta = \frac{\sqrt{3}}{2}$

21. ______________

22. ______________

Name:
Instructor:

Date:
Section:

23. $\tan\theta = \sqrt{3}$

24. $\sin\theta = \frac{1}{\sqrt{2}}$

23. ____________

24. ____________

25. $\cos\theta = \frac{\sqrt{2}}{2}$

25. ____________

For #26-28, a ramp at a skate park has a 12% grade. The ramp is 10 ft long.

26. Draw a diagram of the situation.

27. What angle does the ramp make with the horizontal?

28. How much does the elevation change from one end of the ramp to the other?

27. ____________

28. ____________

Concept Connections

29. Explain a percent grade of a road. Give an example.

____

____

30. What is the difference between $\frac{1}{\sin x}$ and $\sin^{-1} x$?

____

____

Name: Date:
Instructor: Section:

Chapter 6 INTRODUCTION TO THE TRIGONOMETRIC FUNCTIONS

Activity 6.4

Learning Objectives

1. Determine the measure of all sides and all angles of a right triangle.

Practice Exercises

For #1-3, use the following triangle to solve. Round to the nearest tenth.

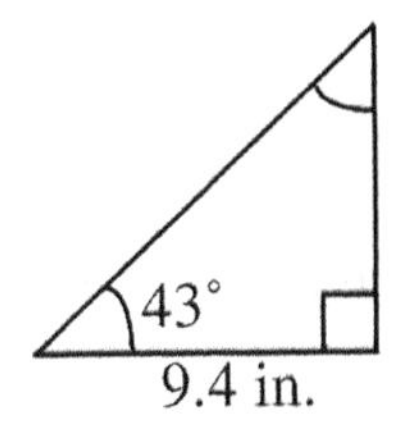

1. Find the hypotenuse. **1.** ____________

2. ____________

2. Find the side opposite to the 43° angle.

3. Find the other acute angle. **3.** ____________

For #4-6, use the following triangle to solve. Round to the nearest tenth.

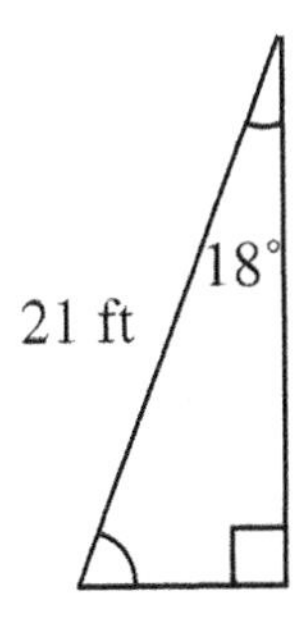

4. Find the other acute angle. **4.** ____________

5. ____________

5. Find the side adjacent to the 18° angle.

6. Find the side opposite to the 18° angle. **6.** ____________

For #7-9, use the following triangle to solve. Round to the nearest tenth.

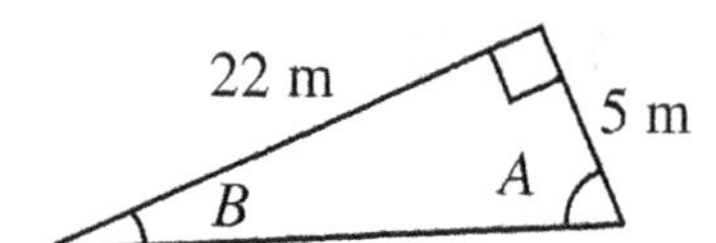

7. Find angle A.

7. ____________

8. ____________

8. Find angle B.

9. Find the hypotenuse.

9. ____________

For #10-12, use the following triangle to solve. Round to the nearest tenth.

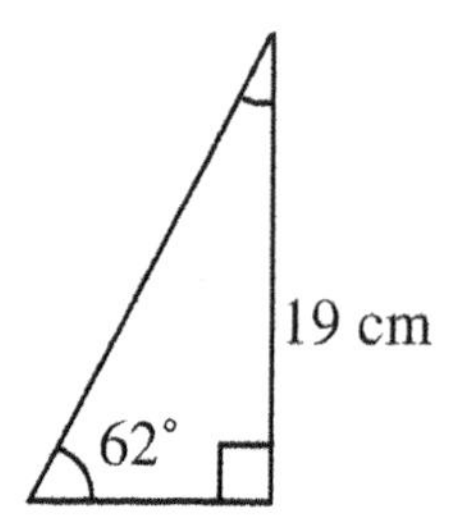

10. Find the other acute angle.

10. ____________

11. ____________

11. Find the side adjacent to the 62° angle.

12. Find the hypotenuse.

12. ____________

For #13-15, use the following triangle to solve. Round to the nearest tenth.

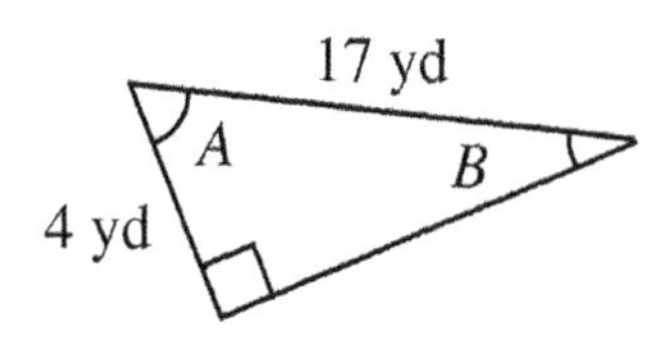

13. Find angle A.

13. ____________

14. ____________

14. Find angle B.

15. Find the side opposite to angle A.

15. ____________

Name: Date:
Instructor: Section:

For #16-18, use the following triangle to solve. Round to the nearest tenth.

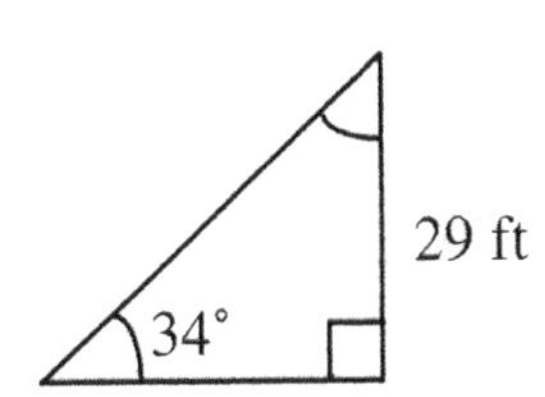

16. Find the other acute angle. **16.** ______________

17. ______________

17. Find the side adjacent to the 34° angle.

18. Find the hypotenuse. **18.** ______________

For #19-21, use the following triangle to solve. Round to the nearest tenth.

19. Find angle A. **19.** ______________

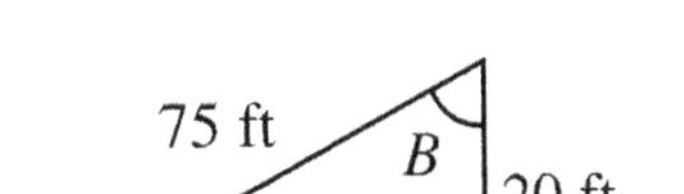

20. ______________

20. Find angle B.

21. Find the side adjacent to angle A. **21.** ______________

For #22-24, use the following triangle to solve. Round to the nearest tenth.

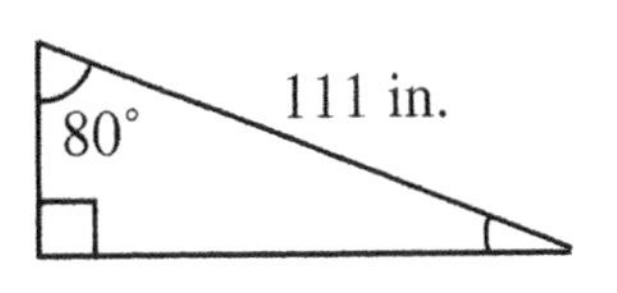

22. Find the other acute angle. **22.** ______________

23. ______________

23. Find the side adjacent to the 80° angle.

24. Find the side opposite to the 80° angle. **24.** ______________

For #7-8, use the following scenario.

A forest ranger knows that a small shack is 150 feet from the base of a steep cliff. Looking from the top of the cliff toward the shack, the angle of depression is 28°.

7. Draw a diagram and indicate the angle of depression.

8. Find the approximate height of the cliff.

8. ______________

For #9-10, use the following scenario.

Looking out from a lighthouse 190 feet above sea level, the angle of depression of a small craft is 10.5°.

9. Draw a diagram and indicate the angle of depression.

10. How far is the craft from the lighthouse?

10. ______________

For #11-13, use the following scenario.

A ladder 36 feet long leans against a wall. It makes an angle of 63.4° with the ground.

11. Draw a diagram and indicate the angle of elevation.

12. Find the vertical distance from the ground to the point where the ladder touches the building.

13. Find the distance from the wall to the point where the ladder touches the ground.

12. ______________

13. ______________

Name: Date:
Instructor: Section:

For #14-15, use the following scenario.
A ramp in a soccer stadium has a horizontal distance of 80 meters and a vertical distance of 15 meters.

14. Draw a diagram and indicate the angle of elevation.

15. Find the angle of elevation of the ramp.

15. ______________

For #16-17, use the following scenario.
On a ship, a sailor observes that the angle of depression of a large rock at the shore is 15.4°. The sailor is 35 feet above the water.

16. Draw a diagram and indicate the angle of depression.

17. Find the distance from the ship to shore.

17. ______________

For #18-19, use the following scenario.
At a particular time of the day, a 60-foot flagpole casts an 80-foot shadow.

18. Draw a diagram and indicate the angle of elevation.

19. Find the angle of elevation of the sun.

19. ______________

For #20-21, use the following scenario.

A copilot of a small aircraft observes a landmark directly below him. The plane maintains an altitude of 2000 ft. A few seconds later the same landmark has an angle of depression of 32°.

20. Draw a diagram and indicate the angle of depression.

21. How far has the plane traveled in the short time?

21. ______________

For #22-23, use the following scenario.

The shadow of a building is 81.7 ft long when the angle of elevation is 35.8°.

22. Draw a diagram and indicate the angle of elevation.

23. Find the height of the building.

23. ______________

For #24-25, use the following scenario.

You are standing 46.6 meters from the base of a tower. Looking toward the top of the tower, the angle of elevation is 37.4°.

24. Draw a diagram and indicate the angle of elevation.

25. Find the height of the tower.

25. ______________

Name: Date:
Instructor: Section:

For #26-27, use the following scenario.
The angle of depression of a tower to a point on the ground 76.5 ft from the bottom of the tower is 28.2°.

26. Draw a diagram and indicate the angle of depression.

27. Find the height of the tower.

27. ______________

Concept Connections

For #28-30, use the following scenario.
The maximum safe angle that a ladder can make with the ground is 65°.

28. What is the shortest ladder a painter will need to paint at a height of 37 feet?

28. ______________

29. Will a fireman be safe in climbing a 54-foot ladder to reach a window that is 48.5 feet above ground?

29. ______________

30. Explain your answer for Exercise #29.

____

____

Name: Date:
Instructor: Section:

Chapter 6 INTRODUCTION TO THE TRIGONOMETRIC FUNCTIONS

Activity 6.6

Learning Objectives

1. Determine the coordinates of points on a unit circle using sine and cosine functions.
2. Sketch the graph of $y = \sin x$ and $y = \cos x$.
3. Identify the properties of the graphs of the sine and cosine functions.

Key Terms

Use the vocabulary terms listed below to complete each statement in Exercises 1–5.

unit circle **central angle** **period** **domain** **range**

1. The shortest horizontal distance it takes for one cycle to be completed is called the ____________________ .

2. A ____________________ has radius 1 and center at (0, 0).

3. The ____________________ of the sine and cosine functions is all angles, both positive and negative.

4. A ____________________ has its vertex at the center of a circle.

5. The ____________________ of the sine and cosine functions is all values of N such that $-1 \leq N \leq 1$.

Practice Exercises

For #6-17, determine the coordinates of the point on the unit circle corresponding to the following central angles. Round to the nearest hundredth.

6. 53° **7.** 145°

6. ________________

7. ________________

8. 315° **9.** –80°

8. ________________

9. ________________

10. 126°

11. 280°

10. ________________

11. ________________

12. 180°

13. 500°

12. ________________

13. ________________

14. –30°

15. 236°

14. ________________

15. ________________

16. 154°

17. 350°

16. ________________

17. ________________

For #18-28, from each of the points on the unit circle, determine the distance traveled from (1, 0) *to the point along the circle. Round to the nearest hundredth.*

18. 53°

19. 145°

18. ________________

19. ________________

20. 315°

21. –80°

20. ________________

21. ________________

Name:
Instructor:

Date:
Section:

22. 126°

23. 280°

22. ______________

23. ______________

24. 180°

25. 500°

24. ______________

25. ______________

26. –30°

27. 236°

26. ______________

27. ______________

28. 154°

28. ______________

Concept Connections

29. Graph the function $y = \sin x$ for $0° \leq x \leq 360°$.

30. Graph the function $y = \cos x$ for $0° \leq x \leq 360°$.

Name: Date:
Instructor: Section:

Chapter 6 INTRODUCTION TO THE TRIGONOMETRIC FUNCTIONS

Activity 6.7

Learning Objectives

1. Convert between degree and radian measure.
2. Identify the period and frequency of a function defined by $y = a \sin(bx)$ or $y = a \cos(bx)$ using the graph.

Practice Exercises

For #1-10, convert from degree measures to radian measures. Give an exact answer and an approximation to 3 decimal places.

1. 30° **2.** 60°

1. ______________

2. ______________

3. 135° **4.** 180°

3. ______________

4. ______________

5. –80° **6.** 171°

5. ______________

6. ______________

7. 270° **8.** 315°

7. ______________

8. ______________

30. How much is 1 radian in degree measure? Given an exact answer and an approximation to the nearest tenth.

Name: Date:
Instructor: Section:

Chapter 6 INTRODUCTION TO THE TRIGONOMETRIC FUNCTIONS

Activity 6.8

Learning Objectives

1. Determine the amplitude of the graph of $y = a\sin(bx)$ and $y = a\cos(bx)$.
2. Determine the period of the graph of $y = a\sin(bx)$ and $y = a\cos(bx)$ using a formula.

Practice Exercises

For #1-10, determine by inspection the amplitude of each function.

1. $y = 13\sin(2x)$ **2.** $y = -2\cos x$

3. $y = 5\cos(3x)$ **4.** $y = -0.8\sin(5x)$

5. $y = 1.9\cos(2\pi x)$ **6.** $y = 3.2\sin x$

7. $y = -8.9\sin(0.5x)$ **8.** $y = \cos(\pi x)$

9. $y = 6\cos(0.5x)$ **10.** $y = 0.5\sin x$

1. ______________

2. ______________

3. ______________

4. ______________

5. ______________

6. ______________

7. ______________

8. ______________

9. ______________

10. ______________

For #11-20, find the period of each function.

11. $y = 13\sin(2x)$

12. $y = -2\cos x$

11. ____________

12. ____________

13. $y = 5\cos(3x)$

14. $y = -0.8\sin(5x)$

13. ____________

14. ____________

15. $y = 1.9\cos(2\pi x)$

16. $y = 3.2\sin x$

15. ____________

16. ____________

17. $y = -8.9\sin(0.5x)$

18. $y = \cos(\pi x)$

17. ____________

18. ____________

19. $y = 6\cos(0.25x)$

20. $y = 0.5\sin x$

19. ____________

20. ____________

For #21-26, for each table, identify a function of the form $y = a \sin bx$ *or* $y = a \cos bx$ *that approximately satisfies the table.*

21.

x	0	0.3927	0.7854	1.1781	1.5708
y	0	−8	0	8	0

21. ____________

Name: Date:
Instructor: Section:

22.

x	0	0.3141	0.6282	0.9423	1.2564
y	4.5	0	−4.5	0	4.5

22. ______________

23.

x	0	0.5236	1.0472	1.5708	2.0944
y	0	7	0	−7	0

23. ______________

24.

x	0	0.1571	0.3142	0.4712	0.6283
y	−8	0	8	0	−8

24. ______________

25.

x	0	0.1963	0.3927	0.5890	0.7854
y	0	2.5	0	−2.5	0

25. ______________

26.

x	0	0.1745	0.3491	0.5236	0.6981
y	−0.6	0	0.6	0	−0.6

26. ______________

For #27-28, match the given equation to one of the graphs that follow. Assume that Xscl = 1 *and* Yscl = 1.

a.

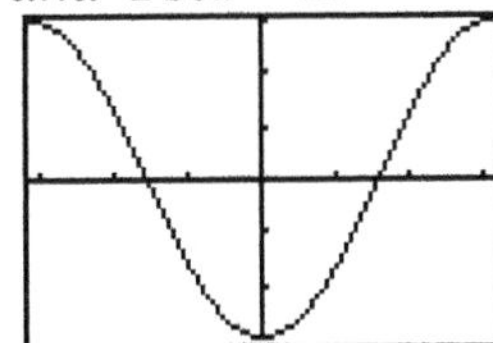

b.

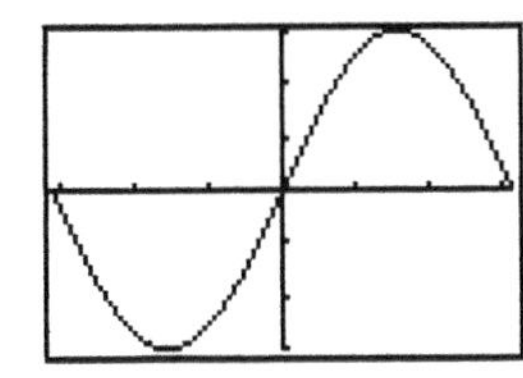

27. $y = 3\sin x$

28. $y = -3\cos x$

27. ______________

28. ______________

Concept Connections

29. Graph the functions $y = \sin x$ and $y = \sin(2x)$ on the same graph using your graphing calculator. What is the period of each function? What is the amplitude of each function? What effect does the number 2 have on the second function as compared to the first?

____

____

30. Graph the functions $y = \cos x$ and $y = 2\cos x$ on the same graph using your graphing calculator. What is the period of each function? What is the amplitude of each function? What effect does the number 2 have on the second function as compared to the first?

____

____

Name: Date:
Instructor: Section:

Chapter 6 INTRODUCTION TO THE TRIGONOMETRIC FUNCTIONS

Activity 6.9

Learning Objectives

1. Determine the displacement of the graph of $y = a\sin(bx + c)$ and $y = a\cos(bx + c)$ using a formula.

Practice Exercises

For #1-3, use the equation $f(x) = \frac{1}{2}\cos\left(\frac{1}{2}x - \frac{1}{2}\right)$.

1. Determine the amplitude.

2. Determine the period.

3. Determine the displacement.

1. ______________

2. ______________

3. ______________

For #4-6, use the equation $f(x) = -8\sin(0.5\pi x + 0.2)$.

4. Determine the amplitude.

5. Determine the period.

6. Determine the displacement.

4. ______________

5. ______________

6. ______________

For #7-9, use the equation $y = 0.3\cos\left(5x - \frac{\pi}{2}\right)$.

7. Determine the amplitude.

8. Determine the period.

7. ______________

8. ______________

9. Determine the displacement.

9. ______________

For #10-12, use the equation $y = 6\sin(x+3)$.

10. Determine the amplitude.

11. Determine the period.

10. ______________

11. ______________

12. Determine the displacement.

12. ______________

For #13-15, use the equation $f(x) = -3.5\cos(4\pi x - 0.1)$.

13. Determine the amplitude.

14. Determine the period.

13. ______________

14. ______________

15. Determine the displacement.

15. ______________

Name: Date:
Instructor: Section:

For #16-18, use the equation $g(x) = 20\sin\left(0.5x + \frac{\pi}{4}\right)$.

16. Determine the amplitude.

17. Determine the period.

18. Determine the displacement.

16. ______________

17. ______________

18. ______________

For #19-21, use the equation $f(x) = -1.7\sin(2x + 1.3)$.

19. Determine the amplitude.

20. Determine the period.

21. Determine the displacement.

19. ______________

20. ______________

21. ______________

For #22-24, use the equation $f(x) = 12\cos(10\pi x + 3)$.

22. Determine the amplitude.

23. Determine the period.

24. Determine the displacement.

22. ______________

23. ______________

24. ______________

For #25-28, match each equation to one of the graphs.

a.

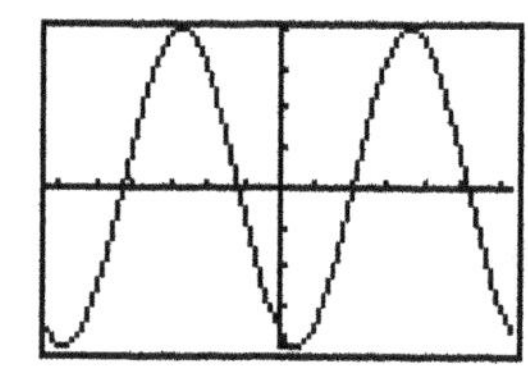

b.

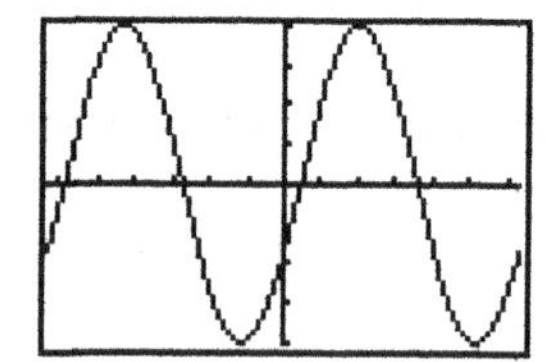

c.

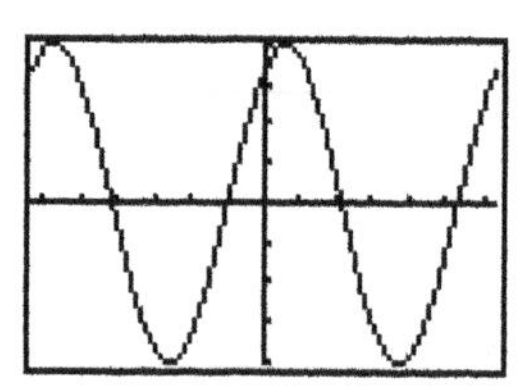

d.

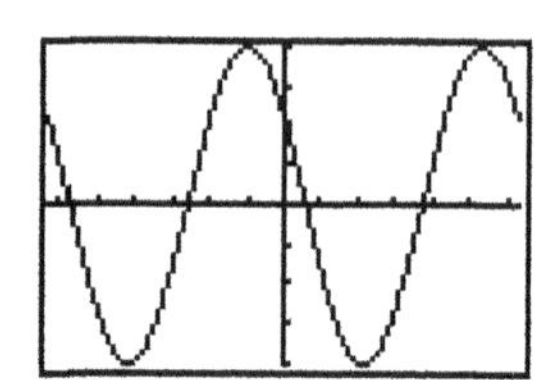

25. $y = 4\sin(x+1)$

26. $y = 4\cos(x+1)$

25. ______________

26. ______________

27. $y = 4\sin(x-2)$

28. $y = 4\cos(x-2)$

27. ______________

28. ______________

Concept Connections

29. Explain what horizontal shift means for the graph of $y = a\cos(bx+c)$, $b > 0$.

____

____

30. Explain the conditions when the horizontal shift moves to the right and when it moves to the left.

____

____

Name: Date:
Instructor: Section:

Chapter 6 INTRODUCTION TO THE TRIGONOMETRIC FUNCTIONS

Activity 6.10

Learning Objectives
1. Determine the equation of a sine function that best fits the given data.
2. Make predictions using a sine regression equation.

Practice Exercises

For #1-5, use the following scenario.

Your share of the heating/cooling bill for 6 consecutive months was recorded in the following table.

Month #	1	2	3	4	5	6
Cost (in $)	10	30	50	50	30	10

1. Determine an appropriate scale, and plot these points.

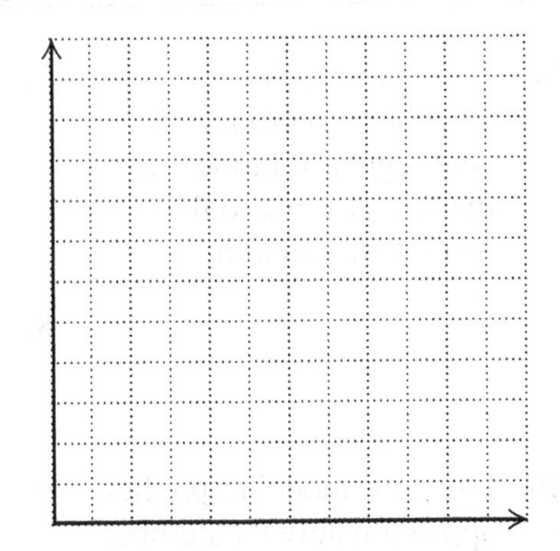

2. Use your graphing calculator to determine the sine regression equation for this data set. Round to the nearest hundredth.

2. ______________

3. Use your graphing calculator to graph the regression equation and data points.

3.

4. Predict the output for month 13.

5. Predict the output for month 17.

4. ______________

5. ______________

For #6-11, use the data listed in the following table, concerning the hours of daylight for the first day of each month for Barrow, Alaska in the year 2000.

Month	1	2	3	4	5	6	7	8	9	10	11	12
Hours	0	4.05	9.20	14.13	19.44	24	24	24	14.45	11.03	5.52	0

6. Determine an appropriate scale, and plot these points.

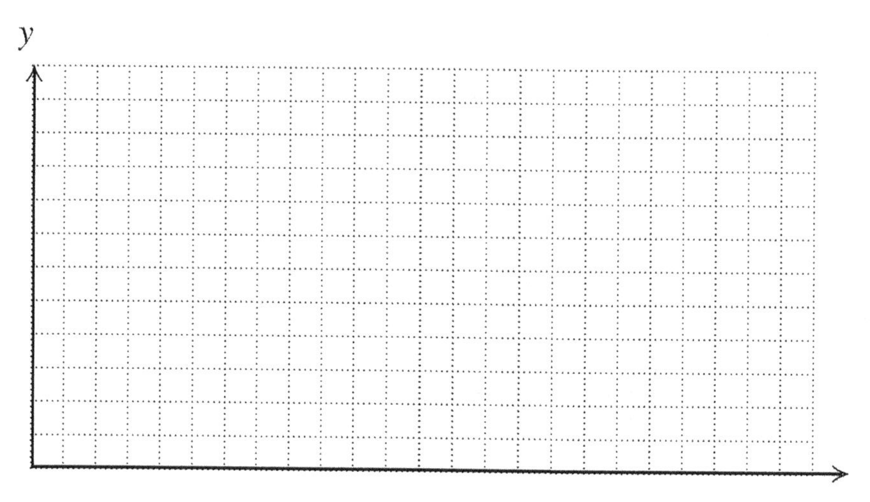

7. Does this data indicate that a sine regression may be appropriate in this case? Explain.

8. Use your graphing calculator to produce a sine regression model for the data. Round to the nearest thousandth.

7. ____________________

____

____

8. ____________________

9. Use your model to predict approximately the number of hours of daylight in mid March.

10. Use your model to predict approximately the number of hours of daylight in April, 2001.

9. ____________________

10. ____________________

11. Do you expect your model in Exercise #8 to be a good predictor of sunlight hours? Explain.

11. ____________________

____

____

Name: Date:
Instructor: Section:

For #12-16, use the data listed in the following table. In training for a marathon, Dan runs for 3 minutes and walks for 1 minute. His heart rate (BPM) was measured.

Minute	5	6	7	8	9	10	11	12	13	14	15
BPM	192	183	175	166	158	151	156	164	171	180	188

12. Determine an appropriate scale, and plot these points.

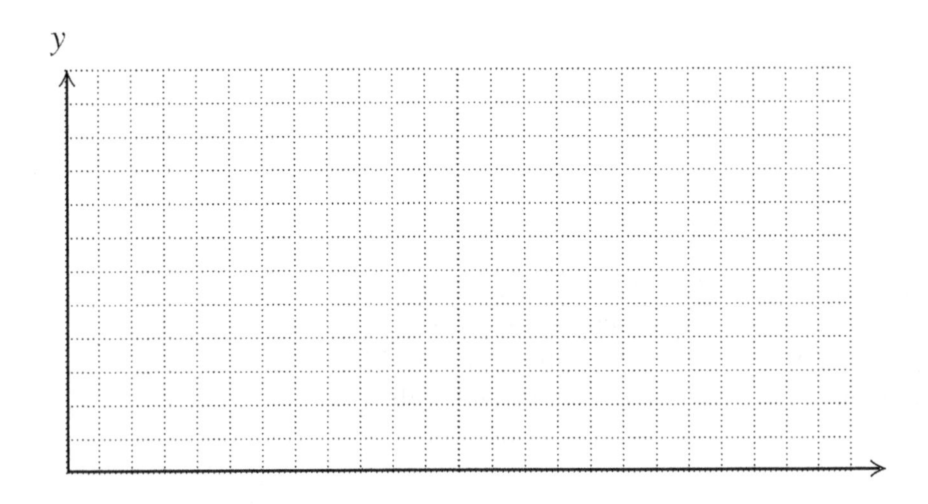

13. Does this data indicate that a sine regression may be appropriate in this case? Explain.

14. Use your graphing calculator to produce a sine regression model for the data. Round to the nearest thousandth.

13. ______________

____

____

14. ______________

15. Use your model to predict approximately the BPM in minute 16.

16. Use your model to predict approximately the BPM in minute 26.

15. ______________

16. ______________

For #17-22, use the data listed in the following table, concerning the number of megawatts used on a hot summer day in California. The 24 hour period starts at midnight.

Hour	0	4	8	12	16	20	24
Megawatts	1295	964	1385	1959	2574	1808	1360

17. Use your graphing calculator to make a scatterplot. Sketch the graph below.

18. Does this data indicate that a sine regression may be appropriate in this case? Explain.

19. Use your graphing calculator to produce a sine regression model for the data. Round to the nearest thousandth.

18. ________________

____

____

19. ________________

20. Use your model to predict approximately the number of megawatts used at 6 PM. Round to the nearest ten.

21. Use your model to predict approximately the number of megawatts used at 8 AM the next day. Round to the nearest ten.

20. ________________

21. ________________

22. Do you expect your model in Exercise #19 to be a good predictor of megawatts used the next day? Explain.

22. ________________

____

____

Name: Date:
Instructor: Section:

For #23-28, use the data listed in the following table, concerning the tide level of Great Channel, Atlantic City. The levels were measured in feet from MLLW (Mean Lower Low Water), every 3 hours.

Hour	0	3	6	9	12	15	18	21	24
Level	0.1	2.0	3.9	2.1	0.2	2.7	5.3	2.4	0.1

23. Use your graphing calculator to make a scatterplot. Sketch the graph below.

24. Does this data indicate that a sine regression may be appropriate in this case? Explain.

25. Use your graphing calculator to produce a sine regression model for the data. Round to the nearest thousandth.

24. ______________

____

____

25. ______________

26. Use your model to predict approximately the tide after 14 hours.

27. Use your model to predict approximately the tide after 30 hours.

26. ______________

27. ______________

28. Use your model to predict approximately the tide after 36 hours.

28. ______________

Concept Connections

29. Do you expect your model in Exercise #25 to be a good predictor of tide height throughout the year? Explain.

____

____

30. Give examples of other real life situations that could be modeled using a sine regression model.

____

____

Odd Answers

Chapter 1 FUNCTION SENSE

Activity 1.1

Key Terms

1. ordered pair
3. variable
5. function

Practice Exercises

7. The output is $h(x)$ or y.
9. y equals h of x.
11. The output is 6.931.
13. g of 7 equals 6.931.
15. The output is 762.
17. 762 equals f of t.
19. The output is salary.
21. Salary is a function of hours, or salary equals s of hours.
23. (6000, 20)
25. Each input has only one output.
27. The input 2 has two different outputs.

Concept Connections

29. Answers may vary. One example is the output is the wages received.

Activity 1.2

Key Terms

1. Dependent

Practice Exercises

3. $f(3) = 9$
5. $f(c) = 5c - 6$
7. $g(-5.1) = -226.7$
9. $h(6) = 11$
11. $h(d) = 11$
13. $p(0.5) = 10$
15. $r(-7) = 20.1$
17. $r(c) = 4 - 2.3c$
19. miles
21. $R(m) = 0.51m$
23. $R(74) = \$37.74$
25. Range: all real numbers
27. Using the domain from Exercise #26, the range is \$0 to \$51.

Concept Connections

29. The domain of the function is the collection of all possible replacement values for the independent variable. The practical domain is the collection of replacement values of the independent variable that makes practical sense in the context of the situation.

Activity 1.3

Key Terms

1. discrete
3. graphically

Practice Exercises

5. $y = -0.5x^2$

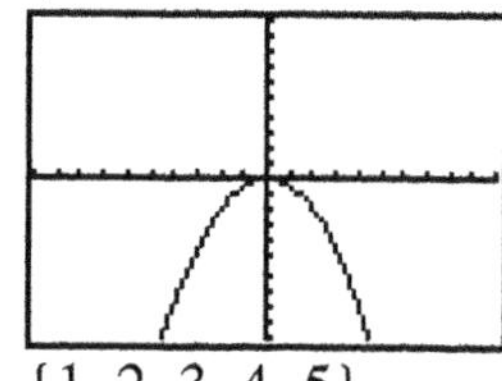

7. $y = 0.5x^2 - 2$

9. $\{1, 2, 3, 4, 5\}$
11. Yes
13. No
15. The employee discount on an item of food is calculated by multiplying the price of the food item by 0.25.
17. Answers may vary.

Item price	4	8	12	16	20
Amount of discount	1	2	3	4	5

19. No
21. $y = 0.0005x^2$
(no graph appears on screen)

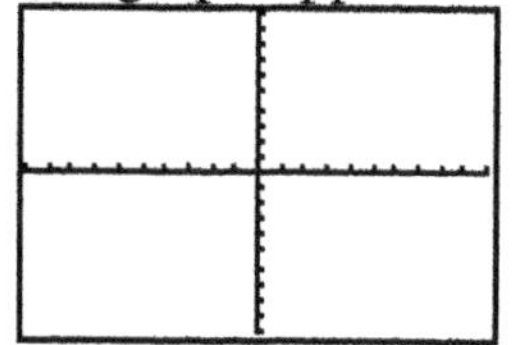

23. The y values are very small. Therefore, you need to have Ymax = 0.01 and Ymin = –0.01.
25. $y = 1000x^2$
(no graph appears on screen)

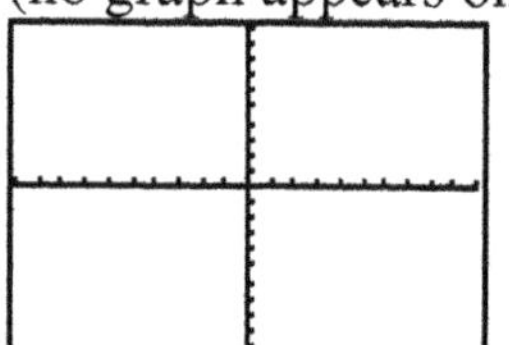

27. The y values are very large. Therefore, you need to have Ymax = 10,000 and Ymin = –10,000.

Concept Connections

29. A discrete function is defined only at isolated input values, and is not defined for input values between those values. A continuous function is defined for all input values, and there are no gaps between any consecutive input values.

Activity 1.4

Key Terms

1. increasing
3. constant

Practice Exercises

5. What is the value of the home after a certain number of years?
7. the value of the home
9.

Independent Variable	1	2	3	4
Dependent Variable	86,250	87,500	88,750	90,000

11. Let v represent the value of the home and t represent the number of years.
13. $v = 85{,}000 + 1250(8) = \$95{,}000$
15. increasing
17. 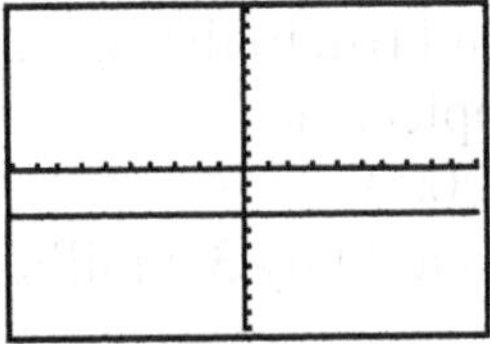
19. The graph is horizontal.
21. decreasing
23. not a function
25. function
27. not a function

Concept Connections

29. A mathematical model can be used to predict output values for input values not in the table of actual data.

Activity 1.5

Key Terms

1. minimum

Practice Exercises

3. The year
5. The account's return is decreasing.
7. The account's return is constant.
9. The account's return is increasing.
11. The account's return is a minimum value.
13. The number of months after origination
15. The percent of car-loan delinquencies increases with time from the origination date.
17. The time in years
19. The population of people living on less than $1.25/day is decreasing with time.
21. The time in years
23. The number of internet users in China increases with time.
25. The time in years
27. The spending decreases until the year 1997; then it increases.

Concept Connections

29. It is a minimum since the line is decreasing from 1989 to 1997 and increasing from 1997 to 2009.

Activity 1.6

Key Terms

1. average rate of change

Practice Exercises

3. 0.789
5. 0.315
7. The national debt is never decreasing over the 30-year time period.
9. 1.126
11. No
13. 0.422
15. The graph would rise to the right.
17. The population is increasing at an average rate of 2.8 million people/year.
19. None
21. From 1990 to 2000
23. During the period from 1990 to 2000, the national population increased by 3.2 million people per year.
25. The national population is not constant during the 100-year time period.
27. Over the 100-year period, the national population is increasing at an average rate of 2.17 million people/year.

Concept Connections

29. The graph for that period remains constant. The graph would be horizontal.

Activity 1.7

Key Terms

1. slope

Practice Exercises

3. Yes
5. No
7. $m = -4$
9. $\left(\frac{1}{4}, 0\right)$
11. $m = -2$
13. $(2, 0)$
15. $m = 3$
17. $(1, 0)$
19. $y = \frac{1}{2}x - 2$
21. $(4, 0)$
23. $(0, 1)$
25. $(1, 0)$
27. The slopes are different; the vertical intercepts are the same, $b = 3$.

Concept Connections

29. The vertical intercept $(0, b)$ of a graph is the point where the graph crosses the vertical axis. The horizontal intercept $(a, 0)$ of a graph is the point where the graph crosses the horizontal axis.

Activity 1.8

Key Terms

1. slope-intercept

Practice Exercises

3. $y=-\frac{2}{3}x+13$
5. $y=\frac{4}{5}x+8$
7. $y=-2x-3$
9. $y=5x+39$
11. $y=-5x+27$
13. $y=\frac{7}{3}x+3$
15. $y=-9x-75$
17. $y=\frac{1}{5}x-12$
19. $y=4x-25$
21. $y=\frac{1}{8}x+58$
23. $y=-x-1.2$
25. $y=10x+3.6$
27. $y=12x-13$

Concept Connections

29. Parallel lines have the same slopes with different y-intercepts.

Activity 1.9

Key Terms

1. vertical

Practice Exercises

3. $y=\frac{4}{5}x-4$
5. (0, –4)
7. $y=\frac{2}{3}x+2$
9. (0, 2)
11. $y=0x-5$
13. (0, –5)
15. $y=-6$
17. (0, –6)
19. yes
21. $x=8$
23. none
25. no
27. Answers will vary. (0, 4), (4, 4), (–3, 4)

Concept Connections

29. Standard form of a linear equation is $Ax + By = C$, where A, B, and C are constants, and A and B are not both zero.

Activity 1.10

Practice Exercises

1.

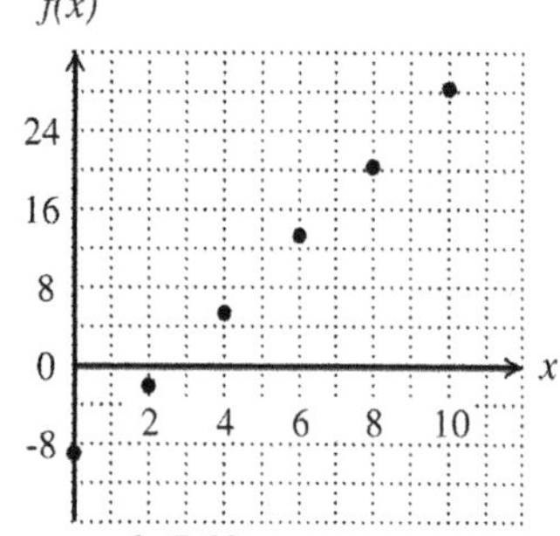

3. $f(x) = 3.763x - 9.414$

5. $m = 3.763$

7. The regression equation yields 9.40.

9. $x = 2.50$

11. Yes, the points are very close to the line.

13. $r = -0.999995$

15. The regression equation yields –4.63.

17. $f(5)$

19.

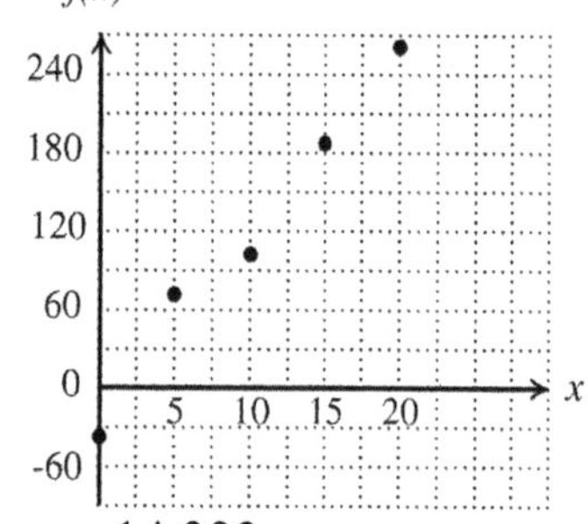

21. $f(x) = 14.322x - 25.464$

23. $m = 14.322$

25. The regression equation yields 332.6.

27. $x = 1.78$

Concept Connections

29. Interpolation uses a regression model to predict an output within the boundaries of the input values of the given data. Extrapolation uses a regression model to predict an output outside the boundaries of the input values of the given data.

Activity 1.11

Key Terms

1. inconsistent

3. dependent

Practice Exercises

5. The answer is (3, 4).

x	y_1	y_2
−2	−1	14
−1	0	12
0	1	10
1	2	8
2	3	6
3	4	4

7. The answer is (0, 5).

x	y_1	y_2
−2	13	3
−1	9	4
0	5	5
1	1	6
2	−3	7
3	−7	8

9. No solution

x	y_1	y_2
−2	4	−1
−1	3	−2
0	2	−3
1	1	−4
2	0	−5
3	−1	−6

11. (–2, 5)

13. (3, 1)
15. (4, 0)
17. (–3, –4.5)
19. (–2, 5)
21. (5, –20.5)
23. $x = -3$
25. $x = 2.9$
27. $x = -8$

Concept Connections

29. If the system is consistent, there is at least one solution, the points of intersection of the graphs. If the system is inconsistent, there is no solution and the lines are parallel. If the system is dependent, there are infinitely many solutions and the equations represent the same line.

Activity 1.12

Practice Exercises

1. $x = 1$
3. $x = 10$
5. $x = \frac{2}{3}$
7. (2, 3)
9. (1, 1)
11. (6, 11)
13. (0, –2)
15. (1, –1)
17. (6, 30)
19. (11, 5)
21. (1, 2)
23. (2, 7)
25. (40, –120)
27. (–3, 2)

Concept Connections

29. Solve one or both equations for a variable. Substitute the expression that represents the variable in one equation for that variable in the other equation. Solve the resulting equation for the remaining variable. Substitute the value from the previous step into one of the original equations, and solve for the other variable.

Activity 1.13

Key Terms

1. inconsistent

Practice Exercises

3. (3, 0, 1)
5. (5, 1, –1)
7. (2, 0, 1)
9. (3, –5, 8)
11. (1, 2, –1)
13. (–2, 4, 1)
15. (7, –3, –4)
17. inconsistent
19. inconsistent
21. yes
23. yes
25. no
27. no

Concept Connections

29. A 3×3 system of linear equations consists of three equations with a total of three variables.

Activity 1.14

Key Terms

1. Elementary row operations
3. reduced row echelon form

Practice Exercises

5. $\begin{bmatrix} 8 & 4 & 3 \\ 2 & -3 & 1 \end{bmatrix}$

7. $\begin{bmatrix} 4 & -3 & -1 & 1 \\ 2 & 2 & 1 & 5 \\ 8 & -1 & 1 & 5 \end{bmatrix}$

9. $\begin{bmatrix} 7 & -3 & -5 & 14 \\ 0 & 1 & 2 & 2 \\ 1 & 0 & -3 & 6 \end{bmatrix}$

11. $\begin{bmatrix} 1 & 7 & -1 & 8 \\ 0 & 1 & 6 & -2 \\ 0 & 0 & 1 & 1 \end{bmatrix}$

13. $2x-8y=5$
    $x+3y=-1$

15. $x-4y+z=3$
    $2x+y-2z=1$
    $x-2y-5z=10$

17. $x+4y-2z=0$
$y+4z=-5$
$x-3z=0$

19. $(x, y, z)=(-4,\ 3,-2)$

21. $\begin{bmatrix} 1 & 0 & 0 & 2 \\ 0 & 1 & 0 & -1 \\ 0 & 0 & 1 & 4 \end{bmatrix}$

23. $(x, y, z)=(1,-1,\ 2)$

25. $(x, y, z)=(1,\ 3,\ 2)$

27. $(x, y, z)=(-1,\ 2,-2)$

Concept Connections

29. The price of a double cheeseburger is \$2.65. The price of the fries is \$1.35. The price of the large soda is \$1.20.

Activity 1.15

Practice Exercises

1. $(-\infty,\ 14)$

3. $(-2.4,\ 13]$

5. $-5 \leq x < 9$

7. $x \leq 2$

9. $x > 7$

11. $x \leq -5$

13. $x > 3$

15. $x \geq 5$

17. $1 < x < 4$

19. $1 \leq x \leq 6$

21. $x \leq -0.2$

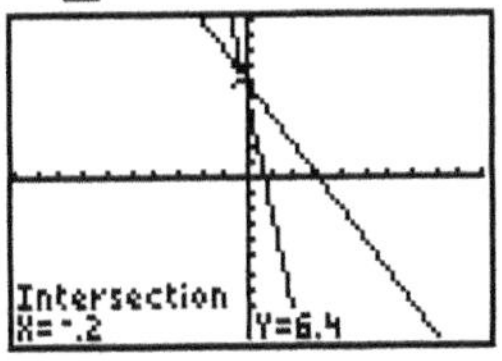

23. $x > -1.5$

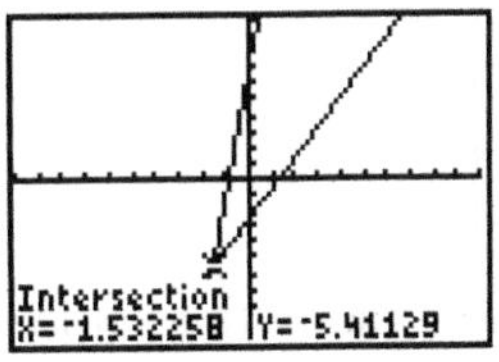

25.
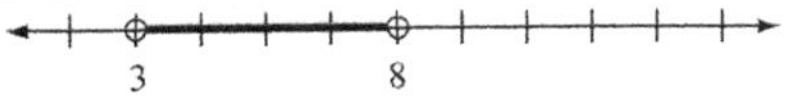

27.
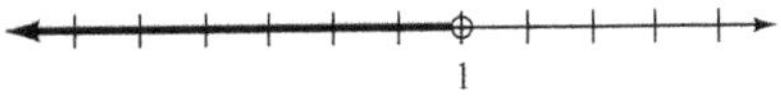

Concept Connections

29. The three approaches are numerical, graphical and algebraic.

Activity 1.16

Practice Exercises

1.

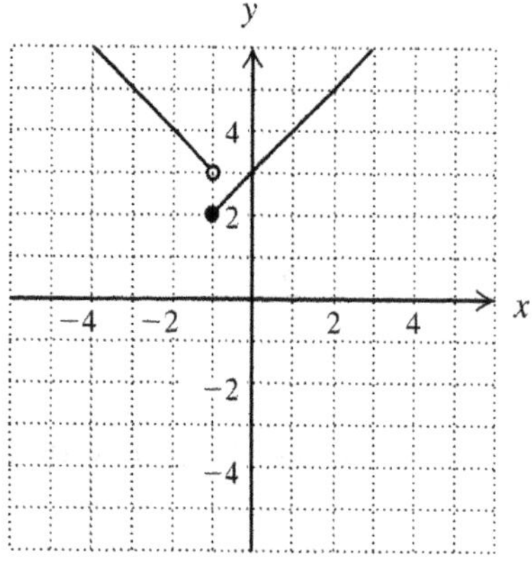

3. $y \geq 2$

5. No solution; 1 is not in the range.

7. All real numbers

9. All real numbers

11. −1

13. 0

15. All real numbers

17.

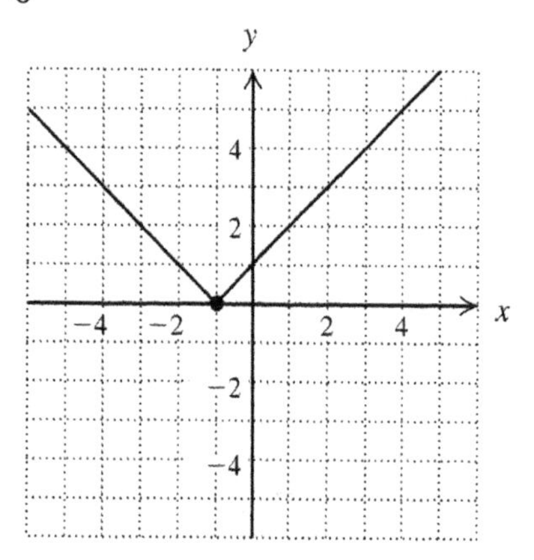

19. $y \geq 0$

21. The graphs are exactly the same.

23.

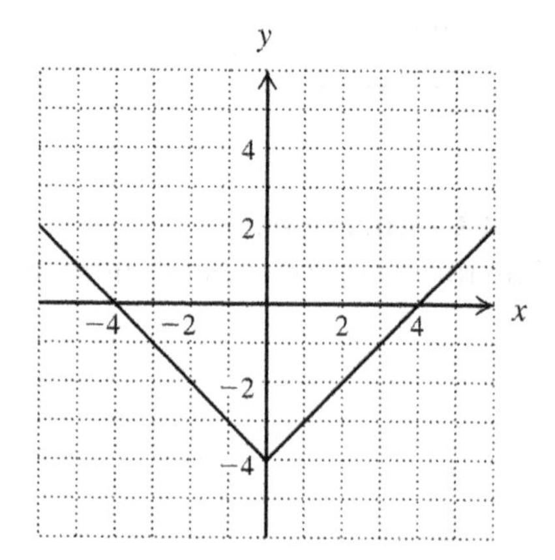

25.

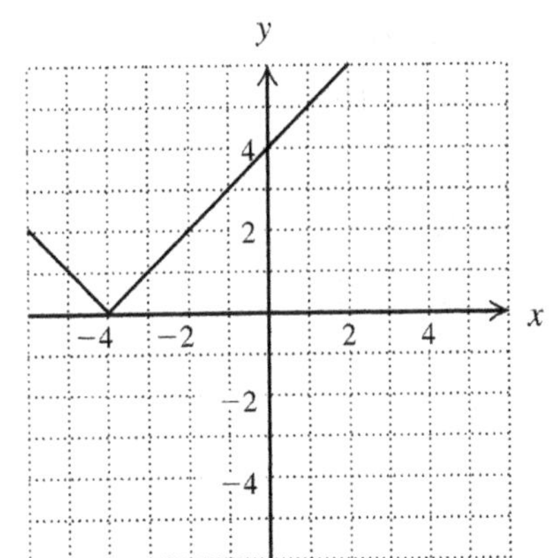

27.

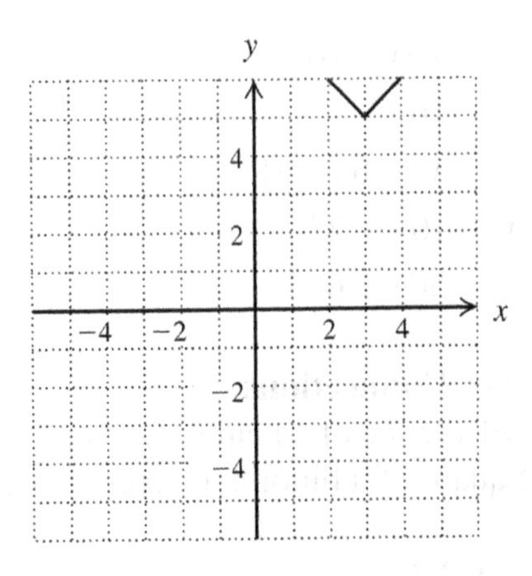

Concept Connections

29. A piecewise function is a function that is defined different for certain "pieces" of the function. The output value is calculated differently depending on the input value.

Chapter 2 THE ALGEBRA OF FUNCTIONS

Activity 2.1

Key Terms

1. trinomial
3. binomial

Practice Exercises

5. no
7. yes, monomial
9. yes, trinomial
11.

x	−1	1	3	5	7	9
$f(x)+g(x)$	−8	−3	4	9	14	19
$f(x)-g(x)$	−4	−3	−4	−3	−2	−1

13. $x^2 - x + 9$
15. $-x^2 + 5x - 3$
17. $-8x^2 + 18x - 5$
19. $-34x - 11$
21. $2x^2 - x$
23. $4x^2 - 4x + 7$
25. $6 - 8 = -2$
27. $3 - 1 = 2$

Concept Connections

29. Paul is right. $\sqrt{3} + x$ is a polynomial, but $\sqrt{3 + x}$ is not.

Activity 2.2

Practice Exercises

1. 5^{11}
3. u^3v^6
5. $24x^{13}$
7. x^{4n}

9. $x^2+10x+16$

11. $x^2+3x-40$

13. x^2-16

15. $4x^2-19x+12$

17. $x^3-x^2-6x+18$

19. x^3-64

21. $x^2-10x+25$

23. $25x^2-16$

25. $x^2-50x+625$

27. $x^2-22x+121$

Concept Connections

29. Difference of squares: Exercises 13, 23, 24, 28.
    Square of a binomial: Exercises 21, 22, 25, 26, 27.

Activity 2.3

Key Terms

1. exponential notation

Practice Exercises

3. 9.6×10^{11}

5. 7.7×10^{14}

7. 1.04×10^{7}

9. 0.000000421

11. 900,000

13. 0.00072

15. 4.0×10^{14}

17. u^3

19. 1

21. $\frac{7}{x^4}$

23. $3w^4$

25. $\frac{6y^7z}{x^3}$

27. $\frac{6c^5}{d^5}$

Concept Connections

29. $4x^0$ is equivalent to 4. $(4x)^0$ is equivalent to 1.

Activity 2.4

Key Terms

1. radicand

3. principal square root

Practice Exercises

5. x^{32}

7. $-125x^{12}$

9. $\frac{36}{w^{14}}$

11. $-a^{40}$

13. 8

15. –8

17. $\frac{1}{25}$

19. $x^{1/15}$

21. $a^{1/2}$

23. $(a-b)^{1/3}$

25. 0

27. –2

Concept Connections

29. Answers will vary. Remind Jeff what $(x^3)^4$ means:
    $(x^3)^4 = x^3 \cdot x^3 \cdot x^3 \cdot x^3 = x^{3+3+3+3} = x^{12}$.
    For y^5y^3, substituting a value for y (other than 0 or 1), will help Jeff check his answer:
    $y^5y^3 = y^{5+3} = y^8$, let y = 2. $2^5 2^3 = 32 \cdot 8 = 256$, and $2^8 = 256$.

Activity 2.5

Practice Exercises

1. 8

3. $f(\text{BOB}) = 3(\text{BOB}) + 2$

5. $f(g(x)) = 3(g(x)) + 2$

7. $g(a) = 2a^2 - 3a + 1$

9. $g(h(x)) = 2(h(x))^2 - 3(h(x)) + 1$

11. $f(g(2)) = 17$

13. $g(f(2)) = 25$

15. $f(g(x)) = 10x - 27$

17. $g(f(x)) = 10x - 9$

19. no

21. $f(g(-1)) = 11$

23. $g(f(-1)) = 71$

25. $f(g(x)) = x$

27. $g(f(x)) = x$

Concept Connections

29. Answers may vary. Using the pairs of functions preceding Exercise #10, or Exercise #15, or Exercise #20, we can see that in general $f(g(x)) \neq g(f(x))$.

Activity 2.6

Practice Exercises

1. $f(x) = 0.06x$

3. The sales tax for a $20.99 dinner is $1.26.

5. $g(20.99) = \$22.25$

7. $h(x) = 0.18(1.06x) = 0.1908x$

9. The customary tip for a $20.99 dinner is $4.00.

11. $k(20.99) = \$26.25$

13. $f(x) = \dfrac{x}{15}$

15. 30 runs are needed to deliver 450 floral arrangements.

17. $g(450) = 10$

19. $P(x) = 3.25x$

21. The florist will pay the drivers a total of $1462.50 for all of the deliveries.

23. $d(450) = 1350$

25. $B(x) = 0.51(3x) = 1.53x$

27. The total mileage expense for the delivery of 450 floral arrangements is $688.50.

Concept Connections

29. The first deal results in a 49% discount. The second deal results in a 44% discount. The first deal is better.

Activity 2.7

Practice Exercises

1. $k^{-1} = \{(15, 0), (12, 1), (-5, 2), (7, 3), (6, 9)\}$
3. $R = \{15, 12, -5, 7, 6\}$
5. $R = \{0, 1, 2, 3, 9\}$
7. $k^{-1}(12) = 1$
9. $k(3) = 7$
11. $k^{-1}(k(3)) = 3$
13. $k^{-1}(k(x)) = x$
15. The cost to rent a car for 3 hours is $40.
17.

Cost	30	35	40	45	50	55	60
Hours rented	1	2	3	4	5	6	7

19. $H(55) = 6$
21. $D = \{30, 35, 40, 45, 50, 55, 60\}$
    $R = \{1, 2, 3, 4, 5, 6, 7\}$
23. $C(7) = 60$
25. $H(C(7)) = H(60) = 7$
27. $H(C(x)) = x$

Concept Connections

29. $f^{-1}(x)$ represents the inverse of the function $f(x)$. $\dfrac{1}{f(x)}$ represents the reciprocal of the function $f(x)$.

Activity 2.8

Practice Exercises

1. $f^{-1}(x) = \dfrac{x+7}{3}$
3. $f^{-1}(a) = \dfrac{5-a}{6}$
5. $h^{-1}(x) = 2x + 1$
7. $g^{-1}(x) = 4 - 3x$
9. $w^{-1}(x) = x - 5$
11. $f^{-1}(x) = \dfrac{x-9}{9}$
13. $h^{-1}(x) = 2x - 2$
15. $g^{-1}(x) = 0.4x + 2$
17. yes
19. no
21. no
23. no
25. $g(5) = 3$

27.

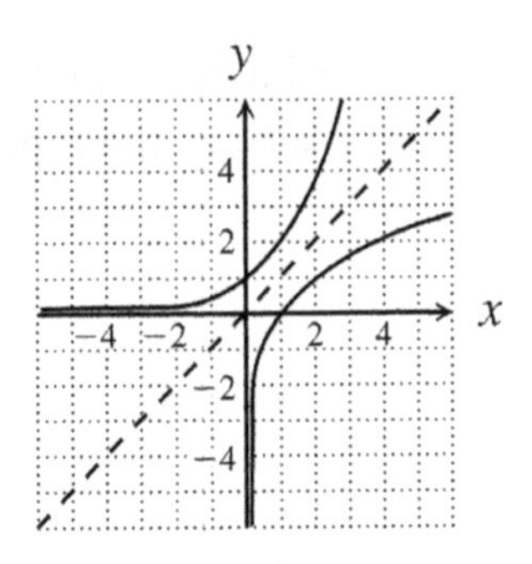

Concept Connections

29. The graphs of inverse functions are reflections about the line $y = x$.

Chapter 3 EXPONENTIAL AND LOGARITHMIC FUNCTIONS

Activity 3.1

Practice Exercises

1.

x	−3	−2	−1	0	1	2	3
$f(x)$	−12	−8	−4	0	4	8	12

3.

x	−3	−2	−1	0	1	2	3
$f(x)$	81	16	1	0	1	16	81

5. Window [–4, 4] [–2, 65]

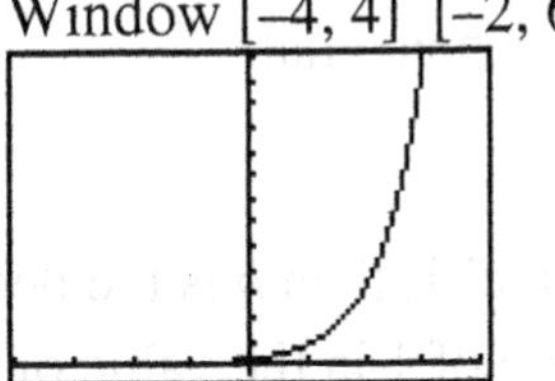

7.

x	−3	−2	−1	0	1	2	3
$f(x)$	54	24	6	0	6	24	54

9.

x	−3	−2	−1	0	1	2	3
$f(x)$	0.005	0.03	0.17	1	6	36	216

11. 6

13. $\frac{9}{5}$

15. Since $0.5 < 1$, it is not a growth factor.

17. This is not an exponential function; there is no growth factor.

19.

x	−3	−2	−1	0	1	2	3
$f(x)$	0.002	0.016	0.125	1	8	64	512

21. $y > 0$

23. 8

25. None

27. The x-axis ($y = 0$)

Concept Connections

29. The horizontal asymptote of the graph of $y = b^x$ is the horizontal axis having equation $y = 0$. The graph of the function gets closer and closer to the x-axis ($y = 0$) as the input gets farther from the origin, in the negative direction.

Activity 3.2

Practice Exercises

1. Window [–10, 4] [–2, 65]

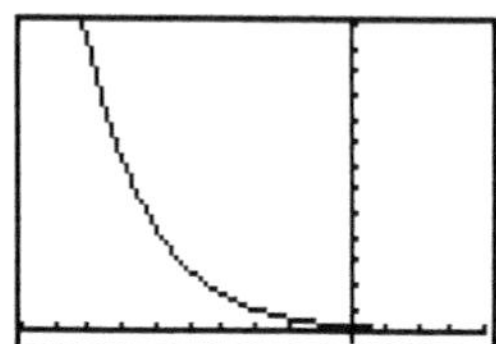

3. The decay factor is $\frac{3}{5}$.

5. All real numbers

7. $y = 0$

9. 0.42

11. Since $\frac{4}{3} > 1$, it is not a decay factor.

13. Since $9.9 > 1$, it is not a decay factor.

15. Since $3.01 > 1$, it is not a decay factor.

17.

x	0	1	2	3
y	100	220	484	1064.8

19.

x	0	1	2	3
y	0.5	0.455	0.414	0.377

21. This data is linear.
23. This data is exponential.
25. This data is linear.
27. This data is exponential.

Concept Connections

29. For an exponential function $y = b^x$, if the base b is between 0 and 1, then b is the decay factor of the function. The graph is decreasing. For each increase of 1 in the value of the input, the output decreases by a factor of b.

Activity 3.3

Practice Exercises

1. (0, 24)
3. $f(0.5) = 41.6$
5. 0.63
7. (0, 8)
9. $g(0.25) = 4.5$
11. 0.30
13. Yes
15. Yes
17. No
19. Yes
21. $f(-2) = 20$
23. $f\left(\frac{1}{2}\right) = 3.54$
25. decreasing
27. increasing

Concept Connections

29. The doubling time of an exponential function is the time it takes for an output to double. The doubling time is determined by the growth factor and remains the same for all output values.

Activity 3.4

Practice Exercises

1. 89%
3. 2.02
5. 29.8%
7. 56%
9. 0.9901
11. 33.24%
13. 37%
15. 4.3%
17. 25.8%
19. 56.2%
21. 28%
23. $V(t) = 50{,}000(0.90)^t$
25. 0.90
27. $V(t) = 50{,}000(0.90)^t$
    Window [–1, 30] [–6000, 60,000]

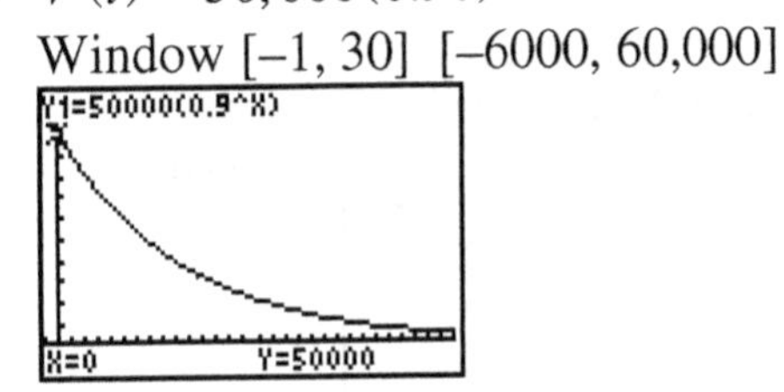

Concept Connections

29. A growth factor, b, is determined from the growth rate, r, by writing r in decimal form and adding 1: $b = 1 + r$.

Activity 3.5

Key Terms

1. continuous
3. compound

Practice Exercises

5. $A = 58{,}000\left(1 + \dfrac{0.035}{4}\right)^{4t}$
7. $82,180.71
9. $6166.77
11. $24,089.67
13. $13,341.85
15. $4146.62
17. 1.510%
19. 7.122%
21. 13.242%
23. 21.341%
25. $A = 5000\left(1 + \dfrac{0.013}{12}\right)^{12t}$
27. In about 53.3 years

Concept Connections

29. Answers may vary. Some examples may include: certificate of deposit, savings bonds, savings account, money market account.

Activity 3.6

Practice Exercises

1. $y = 17e^{0.392t}$
3. 17
5. $y = 14.6e^{-0.821t}$
7. 14.6
9. increasing
11. 7150
13. decreasing
15. 1300
17. increasing
19. 92
21. decreasing
23. 31
25. decreasing
27. 0.737

Concept Connections

29. e is a constant irrational number. The approximate value of e is 2.71828.

Activity 3.7

Practice Exercises

1. Window [–5, 25] [–10, 80]

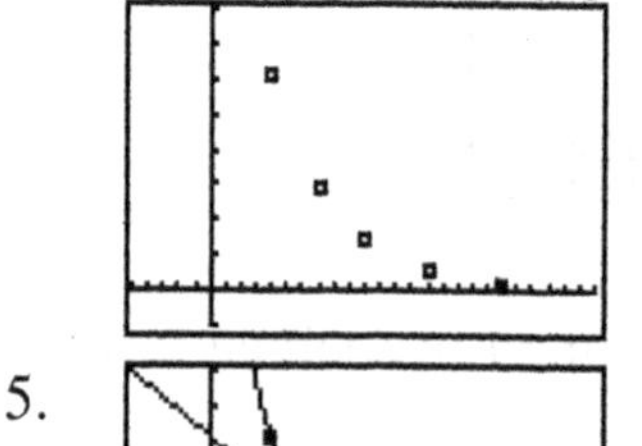

3.

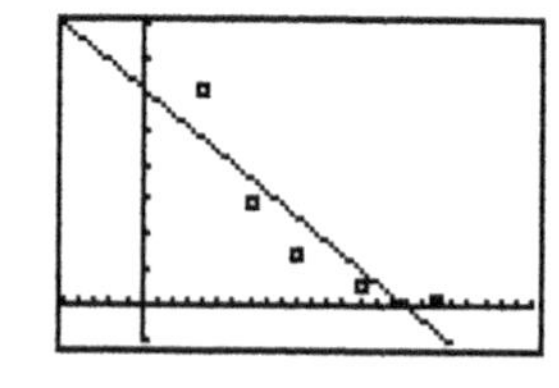

5.

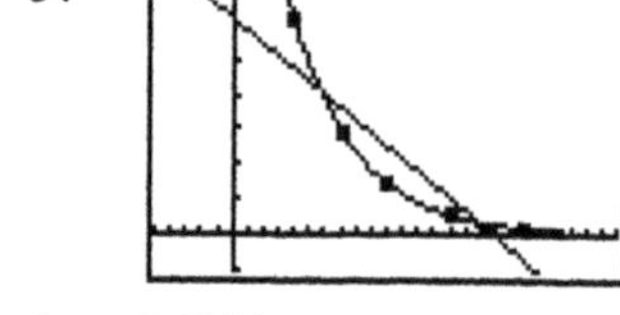

7. 9.0
9. 0.788
11. Yes, the x-axis ($y = 0$)
13. $y = 57.9x - 78.1$
15. $y = 71.4(1.18)^x$
17. The graph of the exponential model is closer to the points on the scatterplot.
19. 1955.9
21. 4.19
23. The set of all real numbers
25. $y = 4(3)^x$ is positive for all values of x.
27. (0, 4)

Concept Connections

29. The intersection is at $x = 16.7$. Window [–5, 25] [0, 50,000]

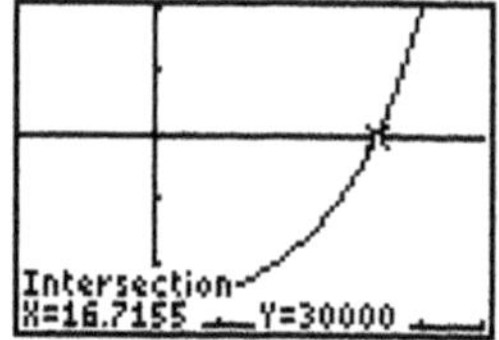

Activity 3.8

Key Terms

1. logarithmic
3. natural

Practice Exercises

5. 6
7. 9
9. –5
11. $\frac{1}{3}$
13. $\frac{2}{3} = \log_{27} 9$
15. $-3 = \ln 0.0498$
17. $5^{-1} = 0.2$
19. $e^{-0.9163} = 0.4$
21. 0.8451
23. 0.9031
25. –0.5229
27. –0.1249

Concept Connections

29. The domain is any positive number. The range is all real numbers.

Activity 3.9

Practice Exercises

1. $y = \log_2 x$
3. $y = \log_{1/2} x$
5. $y = \log_{100} x$
7. $y = 5\ln x$ Window [–4, 25] [–10, 25]

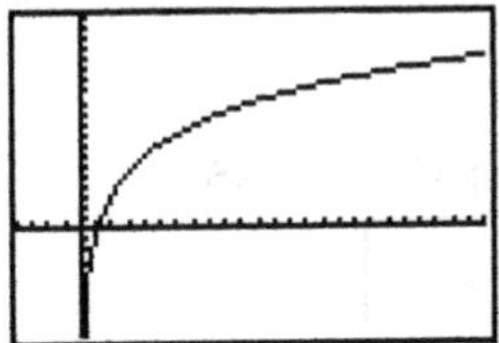

9. Range: all real numbers
11. (1, 0)
13. No, as x increases in value, the output continues to increase slowly. The output values eventually get infinitely large.
15. The output is negative with larger and larger magnitude.
17. $x = -5$
19. $y = \log(5x)$ Window [–4, 25] [–5, 5]

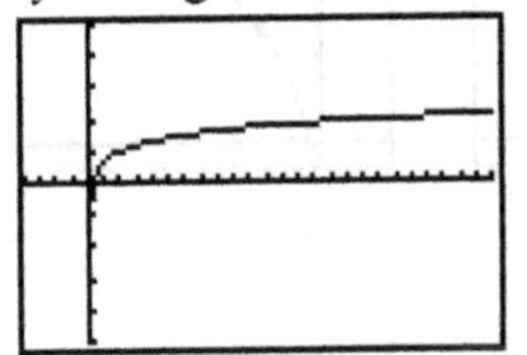

21. Range: all real numbers
23. (0.2, 0)
25.

x	1	0.1	0.01	0.001
y	0.699	−0.301	−1.301	−2.301

27. $x = 0$

Concept Connections

29. The inverse of $f(x) = 10^x$ is $g(x) = \log x$.

Activity 3.10

Practice Exercises

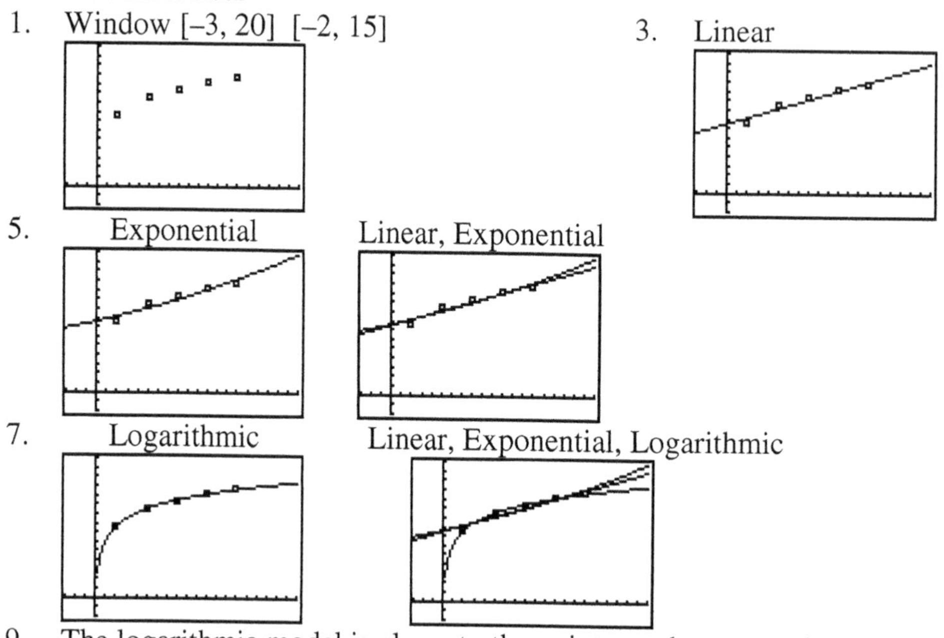

9. The logarithmic model is closer to the points on the scatterplot.
11. $y = 12.3$
13. 0.051
15. The average rate of change decreases as x increases.
17. Logarithmic. It is increasing, but at continually lesser rates.
19. 8.14
21. Window [–2, 13] [–2, 35]

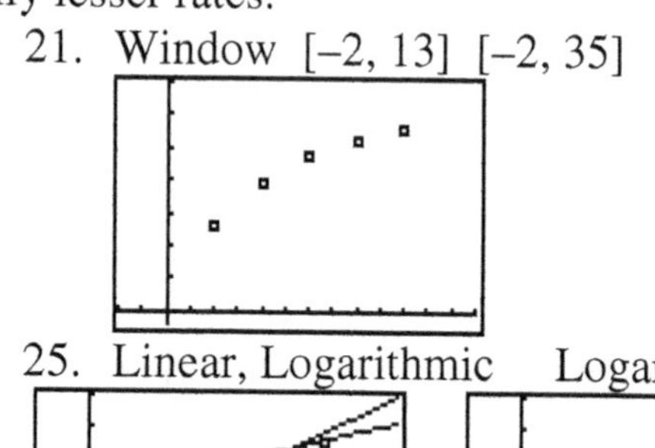

23. Linear

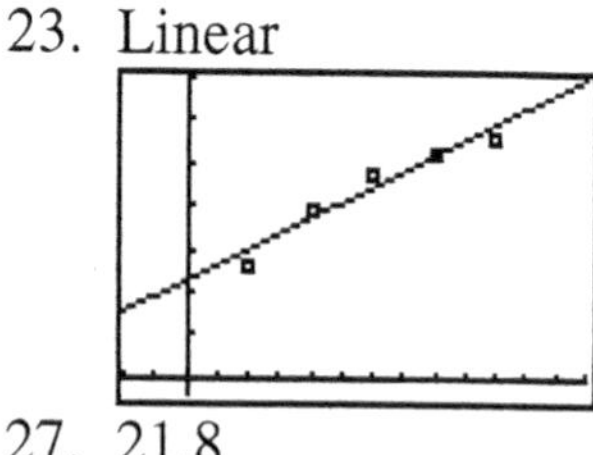

25. Linear, Logarithmic   Logarithmic

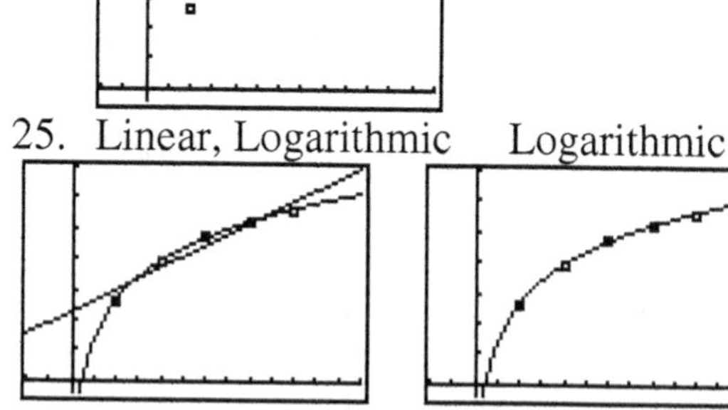

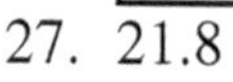

27. 21.8

Concept Connections

29. For an increasing linear function, as the input variable increases, the output increases at a constant rate. For an increasing exponential function, as the input increases, the output increases at an increasing rate. For an increasing logarithmic function, as the input increases, the output increases at a decreasing rate.

Activity 3.11

Practice Exercises

1. $\log_b 6 + \log_b 216$
3. $\log_{13} 11 - \log_{13} 19$
5. $\log_a 90$
7. $\log_a \dfrac{x+1}{11}$
9. 6
11. 41
13. $2\log_b x + \log_b y + 3\log_b z$
15. $x\log_7 7 = x$
17. $\log_3 20$
19. $\ln \dfrac{3^5 x^6}{2^2 3^2} = \ln \dfrac{27x^6}{4}$
21. 3.2047
23. –2.7959
25. 2.2389
27. 1.7829

Concept Connections

29. Answers will vary. Using the rules for a product of logarithms, we have $\log_b x + \log_b y = \log_b (xy)$, not $\log_b (x+y)$.

Activity 3.12

Practice Exercises

1. $x = 3.4594$
3. $t = 5.2983$
5. $x = 1$
7. $x = 2.0959$
9. $x = -18.2754$
11. $x = 0.5$
13. $t = 10.3380$
15. $t = -1.3733$
17. $t = 13.5155$
19. $t = 18.8647$
21. $t = 0.8330$
23. $x = -0.1927$
25. $t = 1386.2944$
27. $t = 3.2981$

Concept Connections

29. The discovery is about 1354 years old.

Chapter 4 QUADRATIC AND HIGHER-ORDER POLYNOMIAL FUNCTIONS

Activity 4.1

Key Terms

1. quadratic

Practice Exercises

3.

x	−3	−2	−1	0	1	2	3
$f(x)$	−18	−8	−2	0	−2	−8	−18

5. $a = 4, b = 0, c = 0$

7. $a = 1, b = -5, c = 0$
9. Since $a = -3 < 0$, the graph is $\cap$-shaped; vertical intercept (0, –7) symmetrical with respect to the y-axis.
11. Both f and g open upward. The low point of f is 7 units below the x-axis; the low point of g is 7 units above the x-axis.
13. Both g and h open upward; h is wider than g; both pass through (0, 0).
15. downward
17. upward
19. upward
21. downward
23. h is wider than g.
25. no
27. no

Concept Connections

29. If $a > 0$, the parabola opens upward. If $a < 0$, the parabola opens downward. The larger the absolute value of a, the narrower the parabola. The smaller the absolute value of a, the wider the parabola.

Activity 4.2

Practice Exercises

1. upward; $x = 0$
3. upward; $x = 2$
5. downward; $x = 4$
7. downward; $x = 0.5$
9. (–5, 2), minimum; (0, 27)
11. (–2.5, 0.75), minimum; (0, 7)
13. (1.25, 0.875), maximum (0, 4)
15. (2, 0), (3, 0); D: all real numbers; R: $g(x) \leq 0.25$
17. (1, 0), (2, 0); D: all real numbers; R: $y \geq -0.5$
19. (–2, 0), (–1, 0); D: all real numbers; R: $y \geq -0.75$
21. (–6, 0), (–1, 0); D: all real numbers; R: $h(x) \leq 6.25$
23. $x > -1; x < -1$
25. $x < 1; x > 1$
27. $x < 1; x > 1$

Concept Connections

29. The axis of symmetry is a vertical line that separates the parabola into two mirror images. The equation of the vertical axis of symmetry is given by $x = \dfrac{-b}{2a}$.

Activity 4.3

Practice Exercises

1. $x = -3$ or $x = 7$
3. $x = 11$ or $x = -4$
5. $x = 5$ or $x = -7$
7. $x = -5$ or $x = 5$
9. $x = -13$ or $x = -2$
11. $x = -4$ or $x = 4$
13. $x = -9$ or $x = 5$
15. $x = 8$ or $x = -2.5$
17. $x = 6$ or $x = -5$
19. $x = 12$ or $x = -6$

21. $x=-8$ or $x=3$

23. $-3<x<7$

25. $-5\leq x\leq 0.75$

27. $x\leq -2$ or $x\geq 8$

Concept Connections

29. There are two different methods. First method: graph $y=f(x)$ and $y=c$. Then determine the points of intersection. Second method: graph $y=f(x)-c$. Then determine the x-intercepts.

Activity 4.4

Practice Exercises

1. $7x^4(5-4x^2)$

3. $3x^2(2x^2-3x+12)$

5. $(x+7)(x-3)$

7. $(x-4y)(x+7y)$

9. $(9+x)(3+x)$

11. $(4x-3)(x+5)$

13. $3b^2(6b-7)(b+5)$

15. $2x^2(2x+3)(4x-5)$

17. $x=5$ or $x=4$

19. $x=7$ or $x=-5$

21. $x=-3$ or $x=\frac{1}{4}$

23. $x=12$ or $x=-9$

25. $x=-\frac{5}{2}$ or $x=-\frac{3}{7}$

27. $x=0$ or $x=\frac{8}{3}$

Concept Connections

29. The Zero-Product Property states that if a and b are any numbers and $a\cdot b=0$, then either a or b, or both, must be equal to zero.

Activity 4.5

Practice Exercises

1. $a=1,\ b=7,\ c=5;$

$$x=\frac{-7\pm\sqrt{7^2-4(1)(5)}}{2(1)}$$

3. $a=6,\ b=16,\ c=-7;$

$$x=\frac{-16\pm\sqrt{16^2-4(6)(-7)}}{2(6)}$$

5. $x=-6.70$ or $x=-0.30$

7. $x=0.70$ or $x=4.30$

9. $x=-3.56$ or $x=0.56$

11. $x=-3.37$ or $x=0.37$

13. $x=-4.5$ or $x=3.5$

15. $x=-5.61$ or $x=1.61$

17. $x=-1$ or $x=-3$

19. $x=-1.56$ or $x=2.56$

21. $x=-3.54$ or $x=2.54$

23. $x=-3.34$ or $x=4.34$

25. $x=-0.63$ or $x=0.75$

27. 3

Concept Connections

29. The quadratic formula is $x=\frac{-b\pm\sqrt{b^2-4ac}}{2a}$, where $a\neq 0$.

Activity 4.6

Practice Exercises

1.

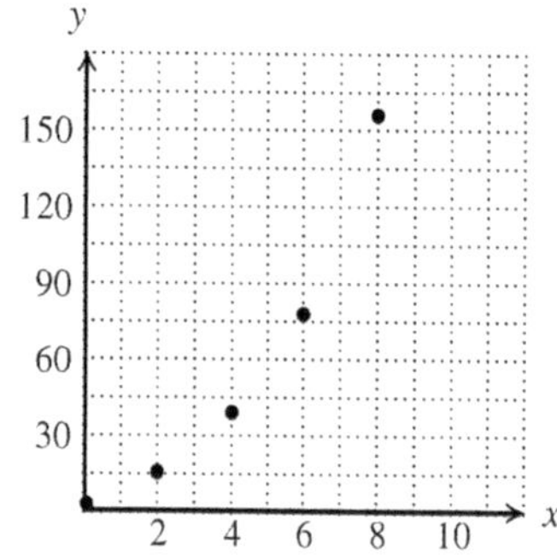

3.

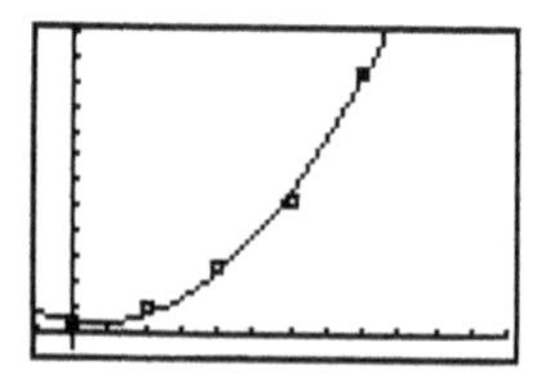

5. 116.776

7. 

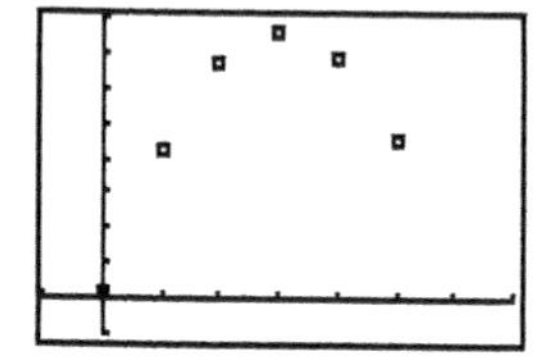

9. Yes; the curve touches nearly every data point.

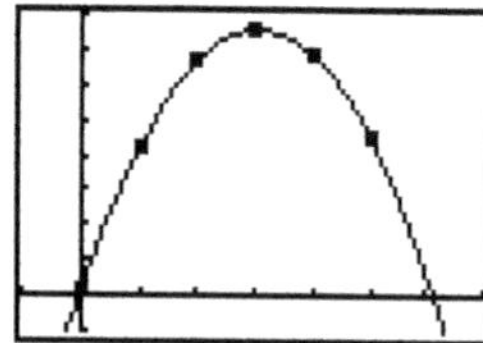

11. All real numbers from 0 to 151 yards.

13. 112 yards

15.

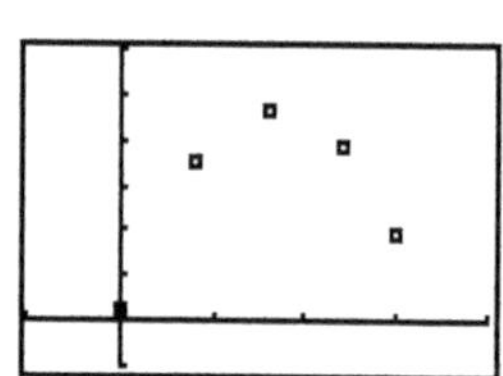

17. Yes; the curve touches nearly every data point.

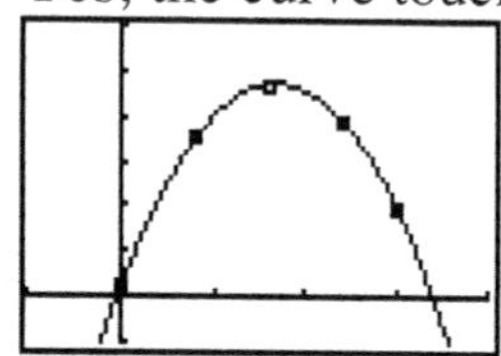

19. All real numbers from 0 to 47 yards.

21. 46 yards

23.

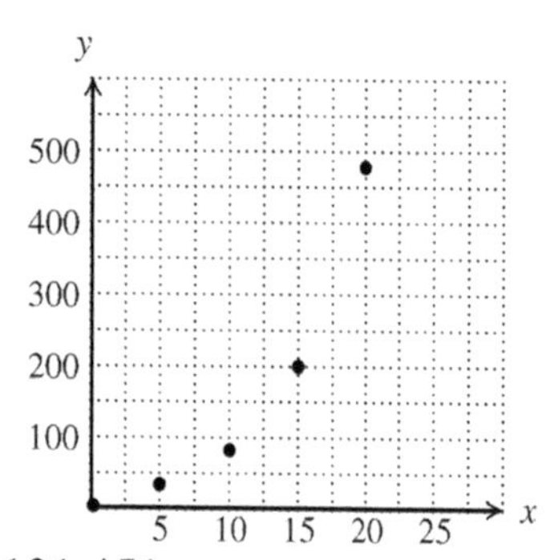

25.

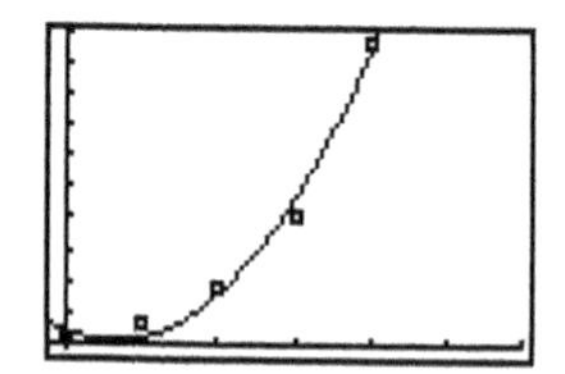

27. 131.451

Concept Connections

29. Parabolic data can be modeled by a quadratic regression equation.

Activity 4.7

Key Terms

1. real
3. imaginary

Practice Exercises

5. $12i$
7. $2i\sqrt{11}$
9. $3i\sqrt{3}$
11. $4i\sqrt{2}$
13. $\frac{7}{8}i$
15. $-2+10i$
17. $9-3i$
19. $19-4i$
21. $x=-2\pm\sqrt{6}$
23. $x=\frac{1}{8}\pm\frac{\sqrt{95}}{8}i$
25. $b^2-4ac=-55$; 2 complex solutions
27. $b^2-4ac=0$; 1 real solution

Concept Connections

29. The graph has no x-intercepts.

Activity 4.8

Key Terms

1. direct

Practice Exercises

3.

x	$\frac{1}{10}$	3	6	9
y	$\frac{1}{2}$	15	30	45

5. $y=5x$
7. $k=\frac{1}{2}$
9. No
11. $k=6$
13. $k=3$
15. 375
17. $k=4$
19. $d=300$
21. Increasing for $x<0$
23. $y=-4x^2$

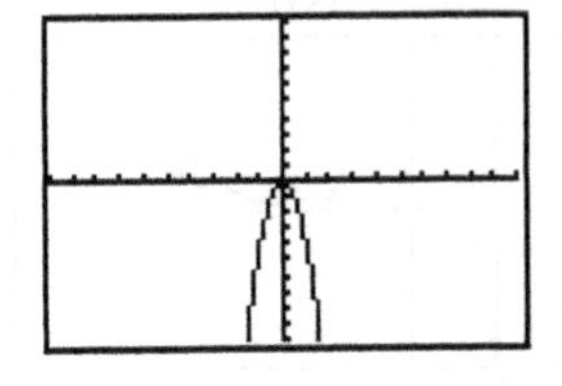

25. $y=\frac{1}{2}x^5$

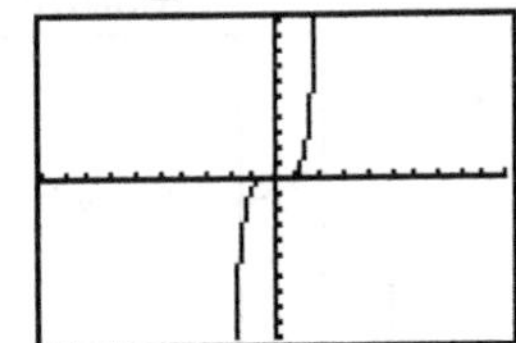

27. $g(x) = 0.1x^3 + 1$

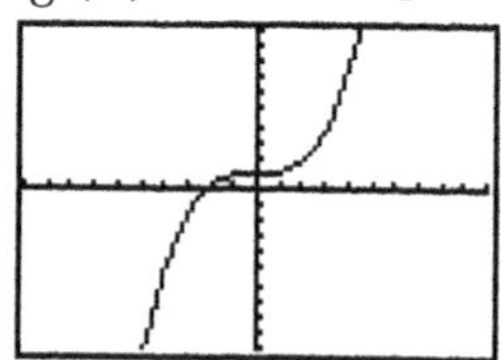

Concept Connections

29. They are also called power functions.

Activity 4.9

Practice Exercises

1. (0, 0), (–2, 0), (3, 0)
3. (0, 0), (1, 0)
5. (–4, 0), (–1, 0), (1, 0), (4, 0)
7. (1, 0), (–1, 0), (2, 0)
9. (–5, 0), (–2, 0), (5, 0)
11. (0, 0), (–3, 0)
13. (0, 0), (–4, 0), (5, 0)
15. degree 4
17. $f(x) = x^4 + 4x^3 - 5x + 4$

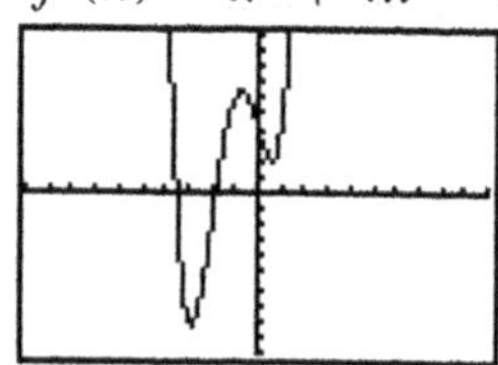

19. y-values greater than or equal to –8.37
21. (0, 4)
23. (–2.85, –8.37), (0.59, 1.99)
25. $-2.85 < x < -0.74$ and $x > 0.59$
27. degree 4

Concept Connections

29. We determine the degree of a polynomial function from the largest exponent on the input variable.

Activity 4.10

Practice Exercises

1. Window: [–5, 25], [–10, 350]

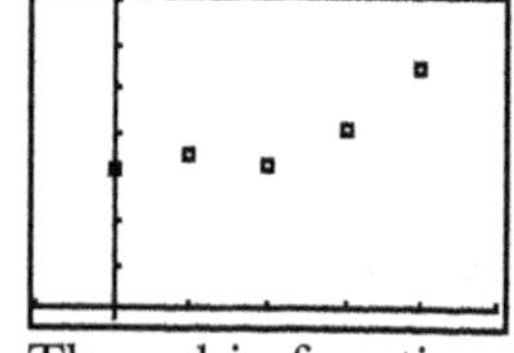

3. $f(x) = 0.48x^2 - 4.08x + 164.23$
5. The cubic function appears best.

Linear

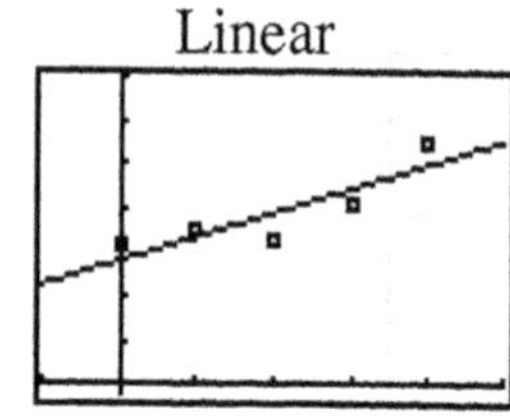

Quadratic

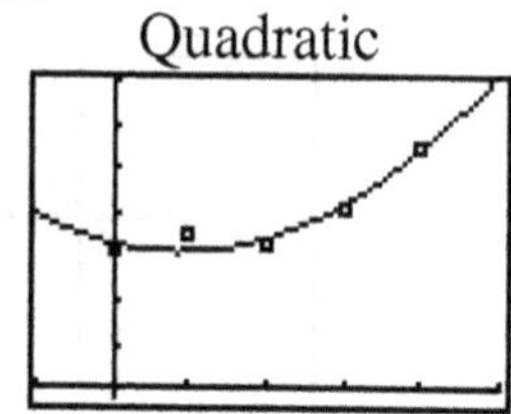

Cubic

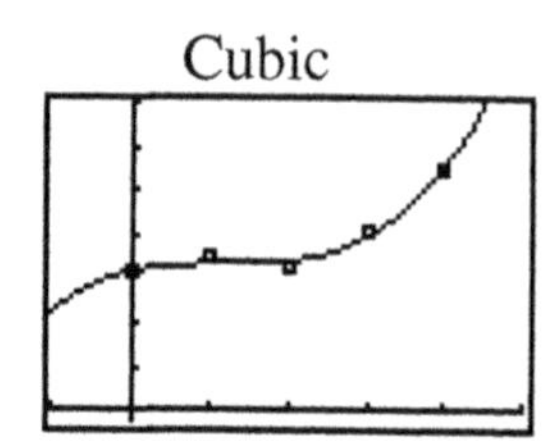

All

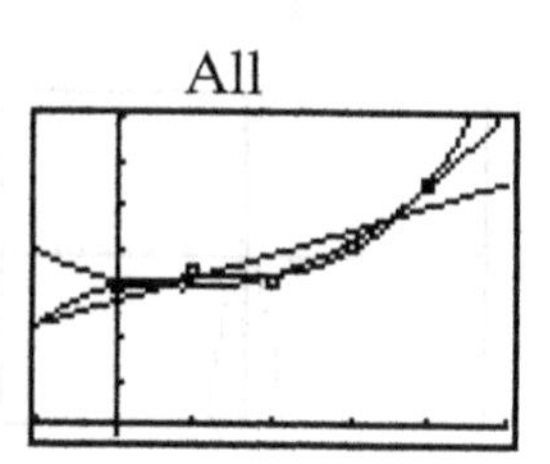

7. $f(7.5) \approx 168$

9. $f(25) \approx 450$

11. Window: [–3, 9], [–50, 120]

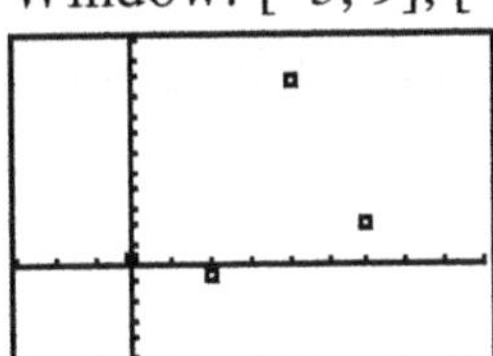

13. $g(x) = -4.38x^2 + 34.27x - 12.76$

15. The cubic function appears best.

Linear

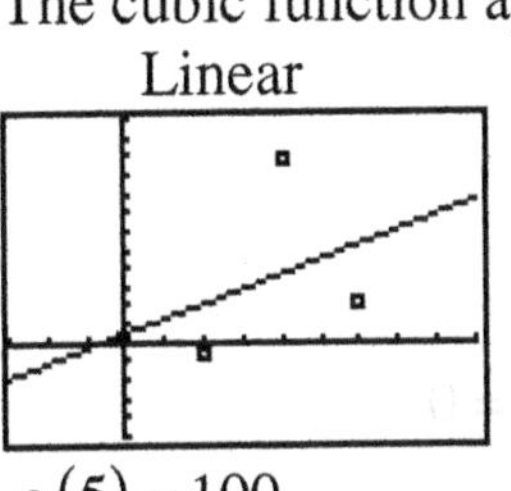

Quadratic

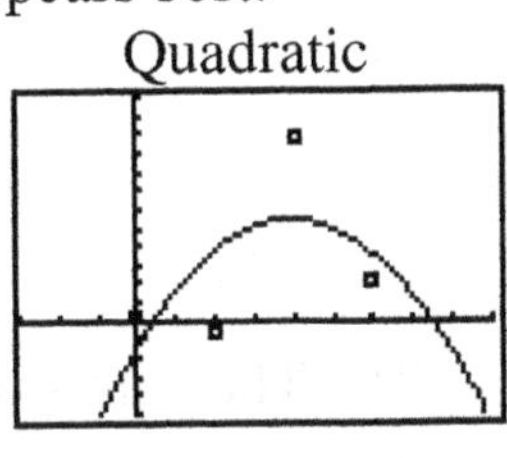

Cubic

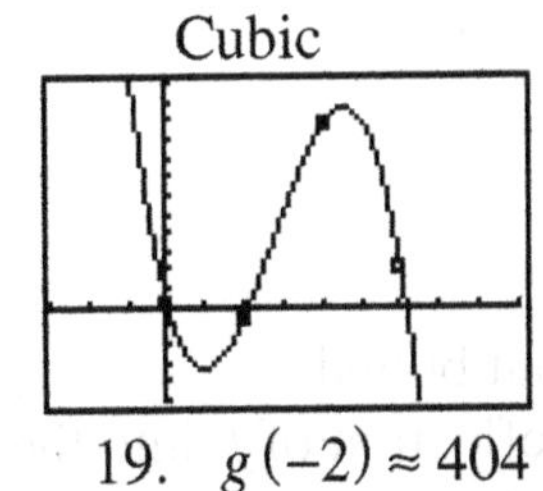

All

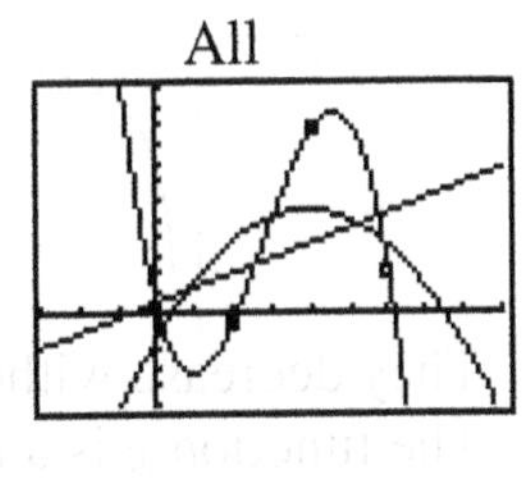

17. $g(5) \approx 100$

19. $g(-2) \approx 404$

21. Window: [–6, 12], [–30, 50]

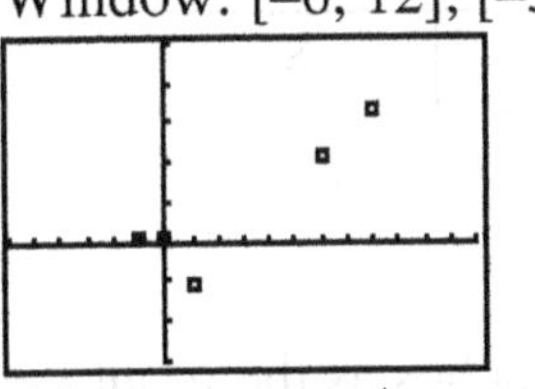

23. $h(x) = 0.82x^2 - 1.77x - 2.78$

25. $h(x) = -0.20x^4 + 2.61x^3 - 6.50x^2 - 8.41x + 1.5$

27. The graph of the quartic equation lies closest to the points, so the quartic equation is the best fit to the data.

Concept Connections

29. The domain is 0 to 20. The range is 156.7 to 277.3.

Chapter 5 RATIONAL AND RADICAL FUNCTIONS

Activity 5.1

Practice Exercises

1. The domain is all nonzero real numbers.

3. $f(x) = \dfrac{3}{x}$

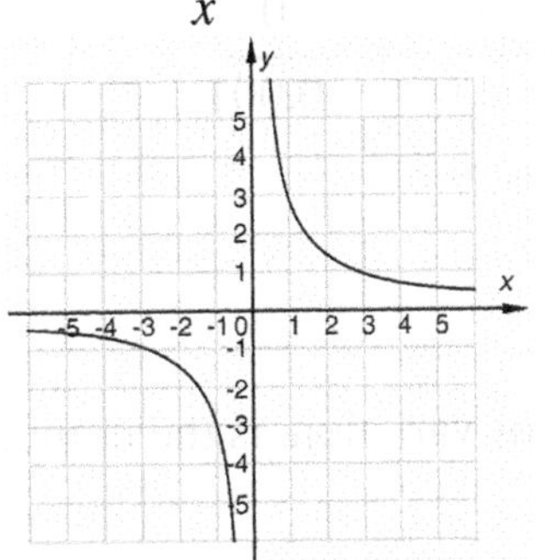

5. They increase, approaching zero.

7. They increase without bound.

9. The y-axis, $x = 0$

11. No, the function is not continuous at $x = 0$.

13. The domain is all nonzero real numbers.

15. $g(x) = -\frac{3}{x}$

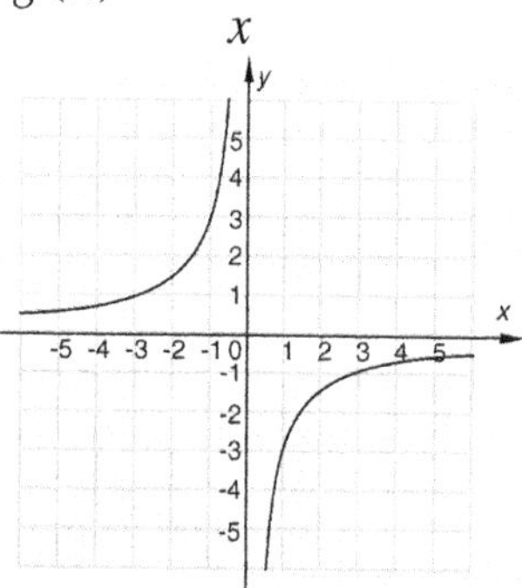

17. They decrease, approaching zero.

19. They decrease without bound.

21. The y-axis, $x = 0$

23. The function g is a reflection of f into the x-axis.

25.

Input, r	50	55	60	65	70	75
Output, t	12	10.9	10	9.2	8.6	8

27. The time increases very rapidly as the rate approaches zero.

Concept Connections

29. The horizontal asymptote is a horizontal line that a graph approaches as the input values get very large in a positive direction (or get very large in a negative direction).

Activity 5.2

Practice Exercises

1. $y = \frac{k}{x^2}$

3. $y = \frac{72}{x^2}$

5. If x is multiplied by 10, the value of y is divided by 100.

7.

x	-10	-5	-1	-0.1	0	0.1	1	5	10
$f(x)$	-0.001	-0.008	-1	-1000	undef	1000	1	0.008	0.001

9. They decrease, approaching zero.

11. The x-axis, $y = 0$

13. They decrease without bound.

15. No

17. The graph is symmetrical with respect to the origin.

19.

x	-10	-5	-1	-0.1	0	0.1	1	5	10
$g(x)$	0.001	0.008	1	1000	undef	-1000	-1	-0.008	-0.001

21. They increase, approaching zero.

23. The x-axis, $y = 0$

25. They increase without bound.

27. No

Concept Connections

29. The constant of proportionality, also known as the constant of variation, is the number k found in inverse variation functions of the form $f(x) = \frac{k}{x^n}$.

Activity 5.3

Practice Exercises

1. all real number except $x = -4$

3. $g(x) = \dfrac{4}{x+4}$

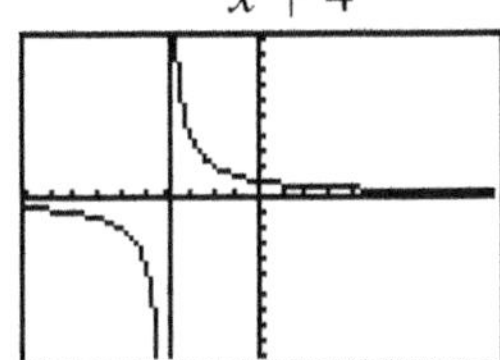

5. As x get larger in magnitude, $g(x)$ gets smaller (closer to zero).

7. yes

9. $x = -4$

11. (0, 1)

13. all real numbers except $x = 13$

15. As x approaches 13, $|f(x)|$ gets very large.

17. $\left(0, -\dfrac{4}{13}\right)$

19. The function $f(x)$ has no maximum.

21. $x = 12$

23. the x-axis, $y = 0$

25. all real numbers except $x = 20$

27. $F(x) = \dfrac{22}{1.5x - 30}$

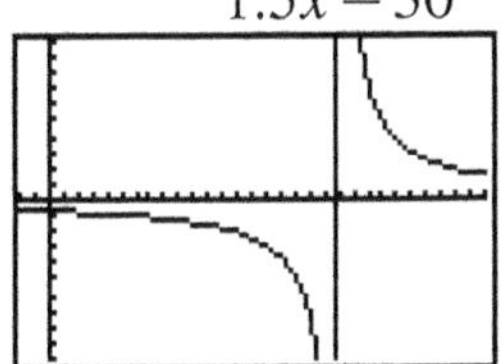

Concept Connections

29. To find the vertical asymptote, set $g(x)$ equal to zero, and solve for x.

Activity 5.4

Practice Exercises

1. $x = 2$

3. $x = \dfrac{3}{4}$

5. $x = 5$

7. $x = -3$

9. $x = 0$

11. $x = 1.25$

13. D: all real numbers except 2
    VA: $x = 2$

15. D: all real numbers except 7
    VA: $x = 7$

17. D: all real numbers except $\dfrac{4}{3}$
    VA: $x = \dfrac{4}{3}$

19. D: all real numbers except $-\dfrac{3}{2}$
    VA: $x = -\dfrac{3}{2}$

21. HA: $y = 3$

23. HA: $y = 1$

25. HA: $y = 3$

27. HA: $y = \frac{1}{2}$

Concept Connections

29. $x = 9.7$

Activity 5.5

Practice Exercises

1. LCD = 42
3. LCD = 135
5. LCD = $8x^5$
7. LCD = $336x^2y^2z^3$
9. LCD = $(x-2)(x+1)$
11. $x = \frac{3}{53} \approx 0.06$
13. $x = \frac{8}{15} \approx 0.53$
15. $x = 10$
17. $x = 2.75$
19. $x = 0.5$
21. $x = -1.85$ or $x = 4.85$
23. $x = 2.4$
25. $z = \frac{xy}{2y+3x}$
27. $c = \frac{5(a+b)}{4}$

Concept Connections

29. Tom takes about 55 minutes. Bill takes about 65 minutes.

Activity 5.6

Practice Exercises

1. $\frac{2}{3}$
3. $\frac{3(x+2)}{x-2}$
5. $\frac{a}{7b^2}$
7. $\frac{15(x-4)}{7(x+4)}$
9. $\frac{3x+5}{x}$
11. $\frac{5x-1}{6}$
13. $\frac{2x^2+7x+1}{(x+2)(x-3)}$
15. $\frac{2-5x}{x}$
17. $\frac{x+22}{20}$
19. $\frac{4x+2}{(x-3)(x+4)}$
21. $\frac{x+3}{3}$
23. $\frac{x}{5}$
25. $\frac{8x+5}{7x+2}$
27. $-(x+7)$

Concept Connections

29. A complex fraction contains fraction in either its numerator or denominator or both.

Activity 5.7

Key Terms

1. radicand

3. radical sign

Practice Exercises

5. 7.07

7. 4.58

9. 64

11. 343

13. $x \geq 6$

15. $x \geq -\frac{3}{5}$

17. $x \leq 5$

19. $x \geq 0$

21. (–5, 0)

23.

25. (0, 5)

27. $(4, 0)$; $(0, -2)$

Concept Connections

29. The graph of $y = -\sqrt{f(x)}$ is a reflection of the graph of $y = \sqrt{f(x)}$ about the x-axis.

Activity 5.8

Practice Exercises

1. $x = 33.64$

3. $x = 36$

5. $x = 27$

7. $x = 49$

9. $x = 4.5$

11. $x = -4$

13. $x = 19$

15. no solution

17. $x = 6$

19. $x = -5$

21. $x = -1$; $x = 2$ does not check.

23. $x = 3.47$

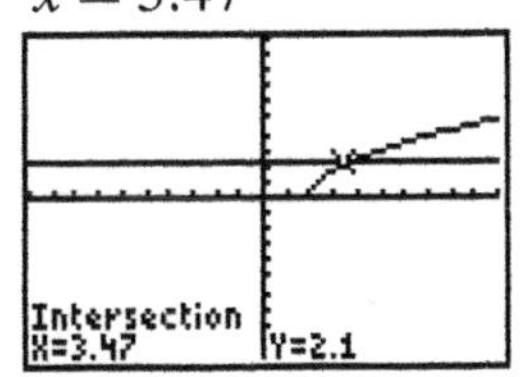

25. $x = -0.70$; $x = -4.30$ does not check.

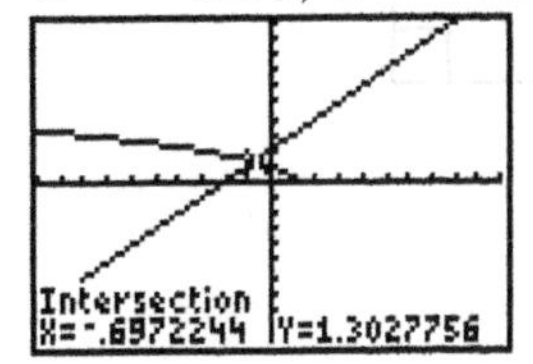

27. 1.8 feet

Concept Connections

29. A radical equation is an equation in which at least one side contains a radical with a variable in the radicand.

Activity 5.9

Practice Exercises

1. 5
3. –6
5. $\frac{1}{12}$
7. 2
9. all real numbers
11. $x \leq 4$
13. all real numbers
15. $x \geq -8.6$
17. $x = 70$
19. $x = -20$
21. $x = -621$
23. $x = -252.8$
25. $x = -13$
27. $x = 343$

Concept Connections

29. The domain of $f(x) = \sqrt[3]{x}$ is all real numbers. The domain of $g(x) = \sqrt[4]{x}$ is $x \geq 0$.

Chapter 6 INTRODUCTION TO THE TRIGONOMETRIC FUNCTIONS

Activity 6.1

Key Terms

1. acute

Practice Exercises

3. 9.3
5. 6.6
7. 111.6
9. 10.5
11. 12.3
13. 1174.8
15. 8.0
17. $\frac{40}{41} = 0.9756$
19. $\frac{9}{41} = 0.2195$
21. $\frac{40}{9} = 4.4444$
23.

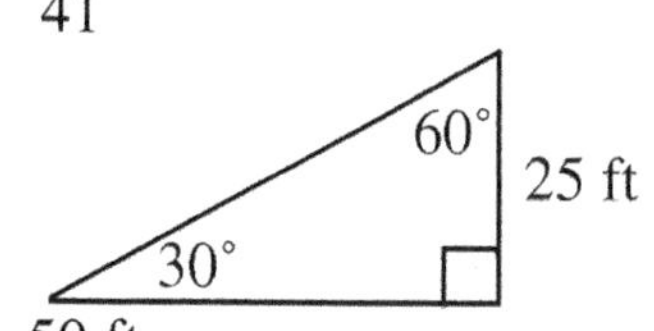

25. 50 ft
27. 7.5 m

Concept Connections

29. Either of the following drawings is correct.

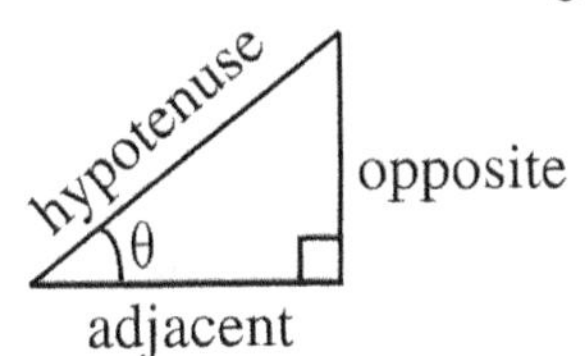

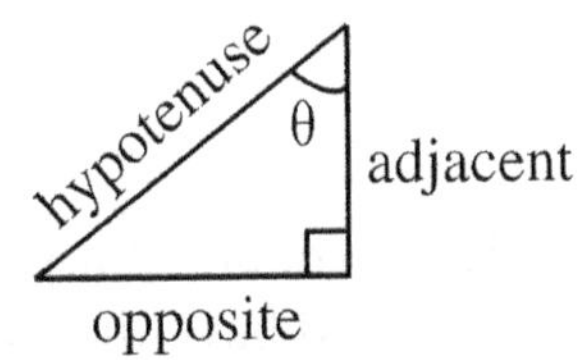

Activity 6.2

Practice Exercises

1. 82°
3. 65°
5. 53°
7. 40°
9. 27°
11. 7°
13. 0.3090
15. 0.5000
17. 0.7431
19. 0.5736
21. 0.5736
23. 0.3256
25. 0.5000
27. 0.6691

Concept Connections

29. Answers will vary.

Activity 6.3

Practice Exercises

1. 45°
3. 67.98°
5. 23.86°
7. 20.04°
9. 63.61°
11. 22.02°
13. 67.98°
15. 55.15°
17. 34.85°
19. 15.05°
21. 30°
23. 60°
25. 45°
27. 6.8°

Concept Connections

29. A percent grade of a road is a term used by highway departments when describing the steepness of a hill. Examples will vary. For example, a steep hill may have an 8% grade.

Activity 6.4

Practice Exercises

1. 12.9 in.
3. 47°
5. 6.5 ft
7. 77.2°
9. 22.6 m
11. 10.1 cm
13. 76.4°
15. 16.5 yd

17. 43.0 ft
19. 15.5°
21. 72.3 ft
23. 19.3 in.
25. 37.6°
27. 93.4 m

Concept Connections

29. No, you can't solve that triangle. There are many similar triangles. For example, triangles with sides 3, 4, 5 and 6, 8, 10 have the same angles.

Activity 6.5

Practice Exercises

1.

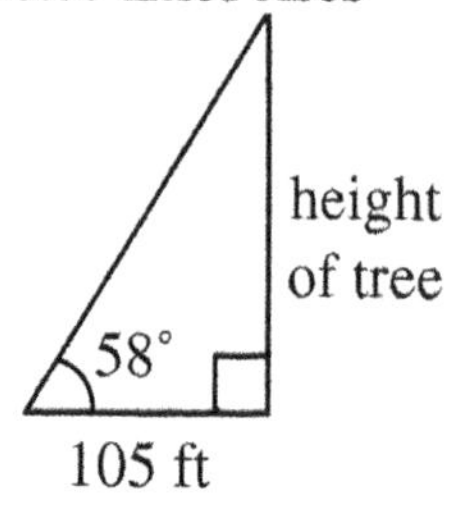

3.

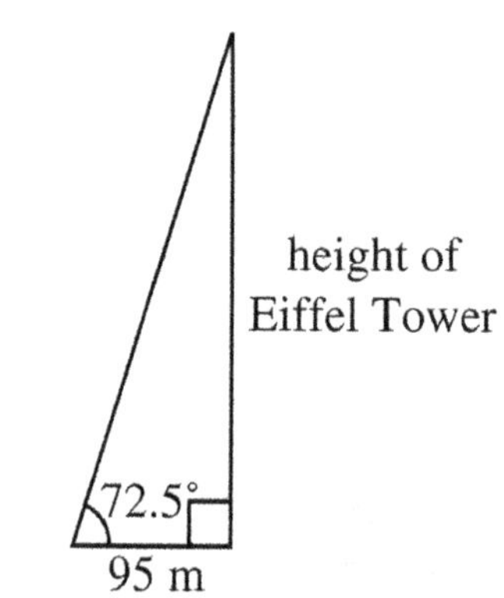

5.

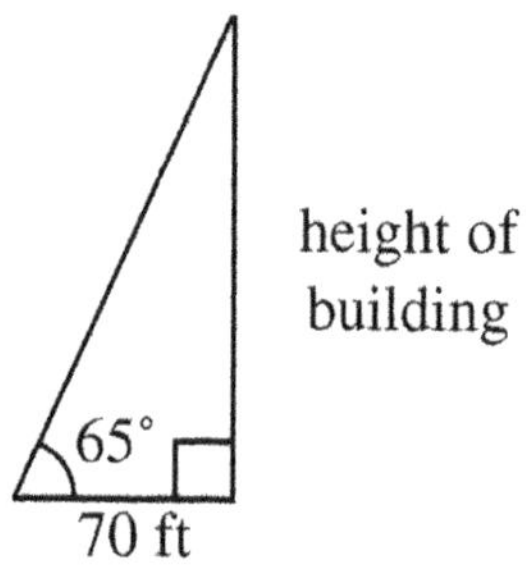

7.

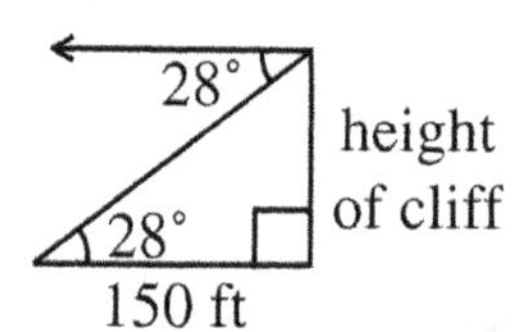

9.

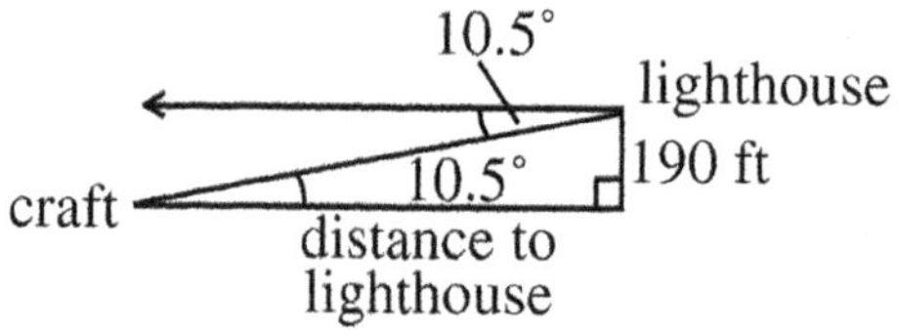

11.

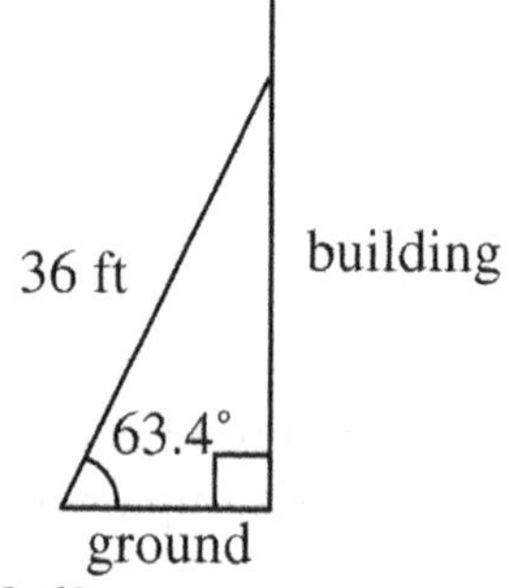

13. 16.1 ft
15. 10.6°
17. 127 ft
19. 37°
21. 3200 ft
23. 58.9 ft
25. 35.6 m
27. 41 ft

Concept Connections

29. Yes.

Activity 6.6

Key Terms

1. period
3. domain
5. range

Practice Exercises

7. (–0.82, 0.57)
9. (0.17, –0.98)
11. (0.17, –0.98)
13. (–0.77, 0.64)
15. (–0.56, –0.83)
17. (0.98, –0.17)
19. 2.53
21. 1.40
23. 4.89
25. 8.73
27. 4.12

Concept Connections

29. Using a graphing calculator, the graph of $y = \sin x$ for $0° \le x \le 360°$ and $-1 \le y \le 1$ is

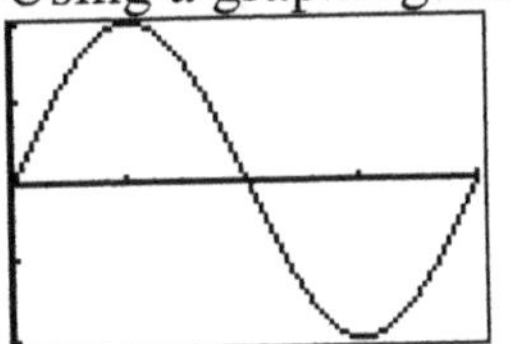

Activity 6.7

Practice Exercises

1. $\frac{\pi}{6} \approx 0.524$ radian
3. $\frac{3\pi}{4} \approx 2.356$ radians
5. $-\frac{4\pi}{9} \approx -1.396$ radians
7. $\frac{3\pi}{2} \approx 4.712$ radians
9. $\frac{5\pi}{4} \approx 3.927$ radians
11. 225°
13. 280°
15. 108°
17. 540°
19. 18°
21. period: π; max: 4.5
23. period: 4π; max 3.7
25. frequency: 2; min: –4.5
27. frequency: 0.5; min: –3.7

Concept Connections

29. The period is the distance between two consecutive maximum values (crests) or two consecutive minimum values (troughs) of the sine function.

Activity 6.8

Practice Exercises

1. 13
3. 5
5. 1.9
7. 8.9
9. 6
11. π

13. $\frac{2\pi}{3}$

15. 1

17. 4π

19. 8π

21. $y = -8\sin(4x)$

23. $y = 7\sin(3x)$

25. $y = 2.5\sin(8x)$

27. b

Concept Connections

29. Using the domain of $-\pi \leq x \leq \pi$, the graphs of $y = \sin x$ and $y = \sin(2x)$ are

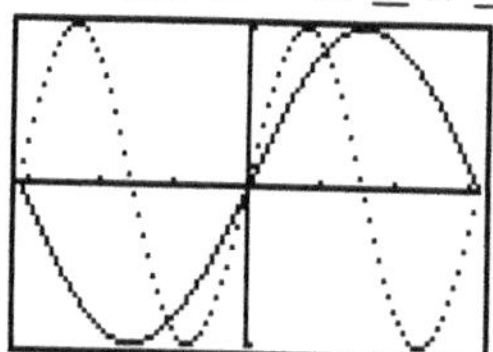

The period of the first function is 2π. The period of the second function is π. The amplitude of both functions is 1. The 2 doubles the number of completed cycles in the second function as compared to the first.

Activity 6.9

Practice Exercises

1. $\frac{1}{2}$

3. 1

5. 4

7. 0.3

9. $\frac{\pi}{10}$

11. 2π

13. 3.5

15. $\frac{0.1}{4\pi} \approx 0.0080$

17. 4π

19. 1.7

21. –0.65

23. $\frac{1}{5}$

25. c

27. a

Concept Connections

29. The horizontal shift, or displacement, is the smallest movement (left or right) necessary for the graph $y = a\cos bx$ to match the graph of $y = a\cos(bx + c)$ exactly. The horizontal shift is given by $-\frac{c}{b}$.

Activity 6.10

Practice Exercises

1.

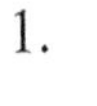

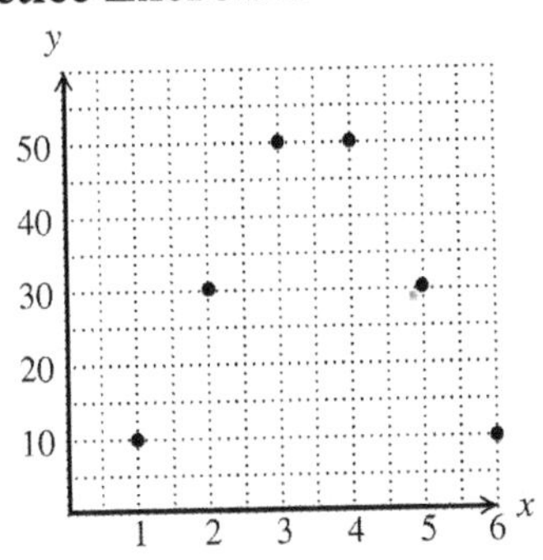

3. Window [–1, 7] [–1, 60]

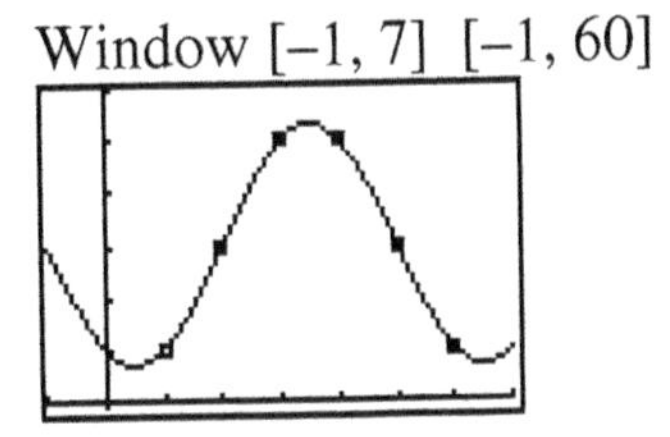

5. \$30
7. Yes. It does show repeated and periodic behavior.
9. $y(3.5) = 11.8$ hours
11. No, adjustments must be made to the number of sunlight hours to account for inclement weather.
13. Yes. It does show repeated and periodic behavior.
15. $y(16) = 192$ BPM
17. Window [–4, 30], [700, 2800]

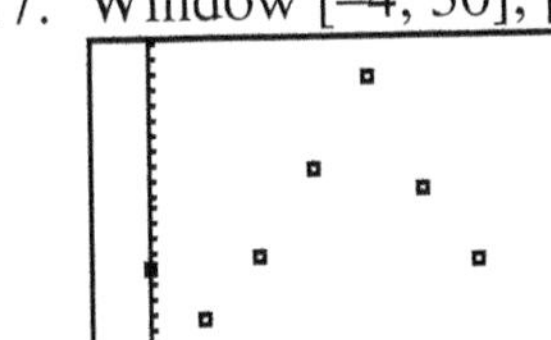

19. $y = 708.545\sin(0.265x - 2.591) + 1680.749$
21. $y(32) = 1410$ megawatts
23. Window [–3, 30] [–3, 8]

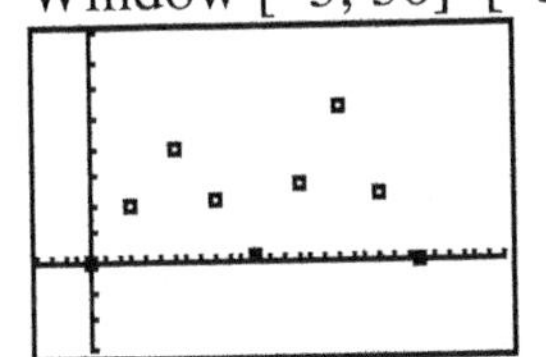

25. $y = 2.236\sin(0.530x - 1.637) + 2.356$
27. $y(30) = 4.6$ feet from MLLW

Concept Connections

29. No, the tide height is affected by weather.